Quick Guide to
BODY FLUID TESTING

Quick Guides in Clinical Laboratory Science

Quick Guide to BODY FLUID TESTING

Second Edition

Darci R. Block

Deanna D. H. Franke

ACADEMIC PRESS

An imprint of Elsevier

Academic Press is an imprint of Elsevier
125 London Wall, London EC2Y 5AS, United Kingdom
525 B Street, Suite 1650, San Diego, CA 92101, United States
50 Hampshire Street, 5th Floor, Cambridge, MA 02139, United States
The Boulevard, Langford Lane, Kidlington, Oxford OX5 1GB, United Kingdom

ISBN 978-0-443-18531-1

For information on all Academic Press publications
visit our website at https://www.elsevier.com/books-and-journals

Publisher: Stacy Masucci
Acquisitions Editor: Michelle Fisher
Editorial Project Manager: Sara Pianavilla
Production Project Manager: Omer Mukthar
Cover Designer: Miles Hitchen

Typeset by STRAIVE, India

Contents

Foreword

Today's medical laboratory is a wonder of technology and automation that would be unimaginable to laboratorians only four or five decades ago. With all the tools now available there may well be the perception that clinical laboratories can perform any analysis requested by a physician or researcher. In some respects, the laboratories of the last century might have actually been better prepared for those tasks since nearly all testing was comprised of what we now call laboratory-developed tests. While accuracy and precision of today's laboratories are demonstrably superior to those of even the recent past, much of the technology has been geared to the analysis of "conventional" biological matrices—largely confined to whole blood, plasma, serum, and urine. Alteration of the sample type can well lead to unpredictable results.

Yet there are a number of situations where analysis of extravascular body fluids provides demonstrated, or at least promising, application. It is an all-too-common scenario for the laboratory to be presented with a specimen type that is unanticipated. My laboratory once received a coffee cup with a request to analyze the residue for a particular body fluid. A tension often exists between the desire to provide a potentially meaningful result and the considerable effort to do so in a reliable and reproducible manner.

Fortunately, Drs. Block and Franke have prepared this *Quick Guide to Body Fluid Testing* to aid the laboratorian in resolving this dichotomy. Don't in any way confuse the "quick" in this book's title with superficial. Through numerous useful figures, tables, examples, and answers to frequently asked questions it provides a wealth of practical information in considerable depth. This *Guide* covers the bases from preanalytical considerations through the reporting process with particular emphasis on the validation of assays for off-label use and the establishment of reliable reference intervals or medical decision points. The authors also provide useful information for laboratories regarding quality control and proficiency testing requirements from regulatory or accrediting bodies for such testing.

Those familiar with the first edition of this *Guide* will be quick to notice the new publisher and expansion of the sections on request- and result report-form design and instrument replacement as well. The authors are clearly passionate about this topic and have a history of presenting educational lectures and workshops as well as peer-reviewed papers in respected journals on analysis of body fluids. They even provide their email addresses should you require further information!

It is good to have this book in print once again. This new edition reflects the continuing interest and increasing maturity of this topic and the *Quick Guide to Body Fluid Testing* will be a useful resource for laboratorians in their efforts to provide clinically actionable results to healthcare providers.

Robert Rej
School of Public Health,
State University of New York at Albany,
Albany, NY, United States

Preface

Laboratories have always tested—or attempted to test—practically any specimen that comes through the door. Specimens other than blood or urine are commonly not listed in the intended use section of the manufacturer's package insert. Since laboratory accrediting agencies view body fluids as unique specimen types, additional validation beyond what is routinely called for is needed. While method validation is a familiar process to clinical laboratorians, its application to body fluids introduces questions that may make the entire process seem daunting and, at times, insurmountable.

The *Quick Guide to Body Fluid Testing* is intended to provide laboratory directors, managers, and medical laboratory scientists with best practice information on the preanalytic, analytic, and postanalytic phases of body fluid testing. In this newly revised second edition, the frequently asked questions (FAQs) are expanded to help the reader appreciate the rationale behind the recommendations provided. We have also expanded the validation strategy to include recommendations for replacing a previously validated test system for body fluid testing. Expanded suggestions for reporting results and establishing long-term quality assurance programs are included.

Chapter 1, Preanalytic Considerations, includes issues to consider when creating orders for body fluid tests, discussions on how body fluids are collected, how specimens should be labeled, as well as concepts important to maintaining specimen integrity for diagnostic evaluation, including transport and processing prior to testing. Chapter 2, Analytic Validation, describes recommended experiments to conduct as part of a comprehensive body fluid validation. Chapter 3, Postanalytic Considerations, discusses key elements to consider when reporting body fluid test results, selecting quality control material and proficiency testing surveys as well as a discussion on the clinical utility of analytes in a number of body fluid types.

The overall goal of the *Quick Guide* second edition is to create a dialog across the patient care continuum to minimize patient risk and maximize diagnostic information for appropriate management. From the laboratory to physicians, nurses, and other medical professionals our hope is that this book becomes a reliable resource for quality body fluid testing practices. Although this *Quick Guide* illustrates many experiments for a comprehensive body fluid validation, it is intended to be and should be considered best practice recommendations for body fluid validation. This *Quick Guide* is not a guarantee of compliance with any laboratory accreditation body; therefore, it is the responsibility of each institution and laboratory director to be aware of and follow applicable local and federal laws and accreditation requirements when designing validation plans and criteria for accepting the body fluid assay performance.

The authors of this *Quick Guide* are clinical laboratory directors with current responsibilities overseeing body fluid testing. The *Quick Guide* is a compilation of laboratory experience and teaching that both have made in clinical laboratory practice as well as through workshops, webinars, short courses, and brown bag presentations through global scientific and medical professional organizations.

Acknowledgments

Darci R. Block

I am still indebted to my mentors and colleagues who have encouraged me and believed in me since I took my first wobbly steps as a newly minted clinical chemist 10 years ago. I am grateful for the opportunities I've been given to share my experience and passion for body fluid testing through writing (textbooks, publications, articles) and presenting (seminars, webinars, and workshops). I am deeply grateful for this encouragement and support. I also wish to express gratitude to the Central Clinical Laboratory and Clinical Specialty Laboratory at the Mayo Clinic for their tenacity and sense of humor about body fluid validation and associated discussions. As iron sharpens iron, this *Quick Guide* is more than it could ever have been if I'd written it alone, thanks to the input, experience, and excellence Dr. Franke lends. We also wish to acknowledge Elsevier for stepping in and taking a chance by expanding their portfolio to include the former AACC Press Quick Guide format. Lastly, I dedicate this work to my family, especially my husband Adam, whose support has made this project possible.

Deanna D. H. Franke

As I reflect on my career in laboratory medicine over the last 20 plus years, I would like to acknowledge and recognize those mentors and fellow clinical chemists who have helped shape and mold me into the practicing laboratory director that I am today. I am forever indebted to this network of professional colleagues for continuous exchange of ideas and information as well as laboratory technologists and teammates with whom I have had the pleasure of working with. To my colleague and friend Darci, your passion is contagious and experience invaluable. Thank you for your partnership—it's been a pleasure rewriting this second edition with you! To my husband and daughter (TLF and SJF), I wish to extend much love and gratitude for your support and understanding for borrowed time. Thanks for being my rock and foundation and giving me the latitude to continue to impact patient care in positive ways.

Abbreviations

AMR	assay measurement range
CAP	College of American Pathologists
COLA	Commission on Office Laboratory Accreditation
CSF	cerebrospinal fluid
CV	coefficient of variation
DOB	date of birth
EDTA	ethylenediaminetetraacetic acid
FAQ	frequently asked question
ISO	International Organization for Standardization
IVD	in vitro diagnostics
K	potassium
LDH	lactate dehydrogenase
LIS	laboratory information system
LLOQ	lower limit of quantitation
MDP	medical decision point
MR	medical record
NA	not applicable
PID	patient identifiers or patient identification
PT	proficiency testing
QC	quality control
SAAG	serum-ascites albumin gradient
SEAG	serum-effusion albumin gradient
TAT	turnaround time

Preanalytic Considerations

Clinical laboratories receive numerous types of specimens, including blood, urine, and other various body fluids. These specimens arrive to the laboratory in different collection and transport devices, including syringes, cups, blood collection tubes, and secondary containers. Although blood is the most common specimen type used to differentiate states of health, body fluids are unique specimens that can provide more localized diagnostic windows into pathologic processes. They can be tested to diagnose, provide prognoses, or monitor disease. Across the patient value stream, providers, nurses, and medical laboratory scientists alike play a crucial role in ensuring appropriate collection, transport, and processing of body fluid specimens. This chapter reviews key concepts that all healthcare teams should know regarding body fluid test ordering, specimen collection, handling, and processing to ensure quality results.

Body Fluid Test Formulary and Ordering Practices

Body fluid test offerings provided by the clinical laboratory should be vetted through continued dialog and engagement with healthcare teams. Leveraging data informatics tools, analyzing body fluid test order patterns, volumes, and specimen types and sources, is a practical starting point. Through this work, laboratories gain a better understanding of the current landscape of orders and potential gaps. Altogether, using data to drive decisions and partnering with clinical stakeholders creates the necessary buy-in to assist the laboratory in formalizing a body fluid lab test formulary, deciding where body fluid testing should take place (i.e., in-house vs send-out), prioritizing test validation planning to meet clinical needs, as well as potentially curbing clinically unnecessary testing.

Patient-centered care delivery should consider where tests may be performed (Fig. 1.1). This decision is driven by several factors: patient acuity, expected turnaround time for clinical management, specimen stability, laboratory resources needed to validate body fluid testing including the cost to maintain validated methods.

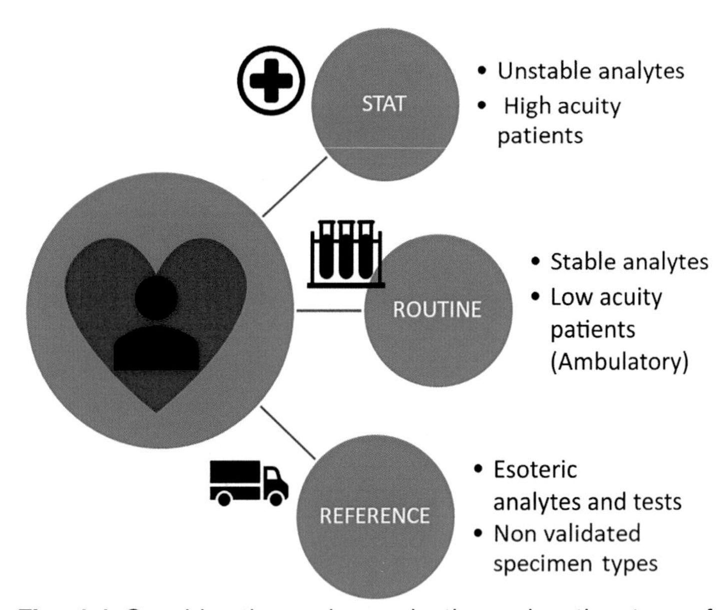

Fig. 1.1 Considerations when selecting a location to perform body fluid testing.

High acuity patients and test analytes with limited stability would best benefit from testing close to the site of collection, for example, pleural fluid pH on blood gas analyzer near the procedural area or within the facility laboratory. Alternatively, specimens with longer stability or test results not needed within 4–8 hours for immediate patient management decisions can more easily be sent to a local routine core laboratory. Consolidating testing to a core laboratory may be beneficial for multisite healthcare systems as this streamlines processing for handling body fluid specimens that may be shared with centralized departments like microbiology and may deliver some cost savings without compromising care delivery. Lastly, more esoteric test requests and body fluid sources that have not been validated by the laboratory could be sent out to a verified reference laboratory.

Governmental incentive programs for implementing electronic medical records and fulfilling meaningful use requirements have propelled healthcare systems and clinical laboratories into electronic order creation. Not surprisingly, these initiatives often move body fluid test ordering from largely paper based to electronic for the first time [1]. Current nomenclature and terminology used in clinical practice such as SNOMED Clinical Terminology and Logical Observation Identifiers Names and Codes (LOINC) are two coding languages that laboratories should consider consulting to standardize procedure and testing nomenclature [2,3].

It is important for clinical laboratories to determine and employ a standardized approach to body fluid test offerings. When publishing a body fluid test formulary, there are several test ordering strategies that may be employed (Fig. 1.2). Common approaches

A Specific

- ☐ Cell Count, Bronchial Alveolar Lavage
- ☐ Cell Count, CSF
- ☐ Cell Count, Pericardial Fluid
- ☐ Cell Count, Peritoneal Dialysate
- ☐ Cell Count, Peritoneal Fluid
- ☐ Cell Count, Peritoneal Lavage
- ☐ Cell Count, Pleural Fluid

B General

- ☐ Cell Count, Body Fluid

with drop down selection

Source:

Bronchial Alveolar Lavage
CSF
Pericardial Fluid
Peritoneal Dialysate
Peritoneal Fluid
Pleural Fluid

and / or with buttons

CSF | Pleural | BAL | Peritoneal
Pericardial | Peritoneal Dialysate

C Order Set – Lumbar Puncture

Common tests

- ☐ Cell Count, CSF
- ☐ Total Protein, CSF
- ☐ Glucose, CSF

Specialized tests

- ☐ Lactate, CSF
- ☐ Culture, CSF
- ☐ Oligoclonal index and banding, CSF

Fig. 1.2 Body fluid test ordering options include offering body fluid source-specific tests (A) general tests that utilize dropdown boxes and/or clickable buttons to specify source (B) or contained within procedure orders, for example CSF tests in a lumbar puncture order set (C).

include general (Total Protein—Body Fluid) versus specific (Total Protein—Pleural Fluid, Total Protein—Pericardial Fluid, Total Protein—Peritoneal Fluid, Total Protein—Synovial Fluid) orderables. It may be as simple as requiring selection of site and source from a specified dropdown list, creating buttons, or using order logic rules to embed orders into body fluid ordersets. Specific orders grouped into a body fluid orderset and/or incorporating body fluid test orders into procedural ordersets (e.g., Paracentesis, Thoracentesis, Lumbar Puncture) may ensure providers have a clear picture of what testing is available guiding proper ordering for the correct source, specimen, and time of collection to facilitate workflows for the performing laboratory. A review of body fluid test offerings from four large reference laboratories indicates a mix of both practices, and, although no answer is definitively correct, both methods have advantages and disadvantages that should be considered as ordering practices are developed and deployed (Table 1.1). These order strategies may be limited by institutional clinical ordering systems in use; therefore, review of electronic order entry tools that are available may also help shape this decision.

Collection

Most body fluids are ultrafiltrates of blood. Body fluids function to support the delivery and removal of nutrients and metabolic by-products from surrounding tissue compartments. Pathogenic processes, including infection, malignancy, and autoimmune and inflammatory diseases, can disrupt the normal production, circulation, and exchange of body fluids, leading to accumulation [4]. Increased volume of fluid in any organ, tissue, or joint compartment usually necessitates clinical intervention to actively remove or drain the accumulated fluid. Altogether, body fluids can be collected for diagnostic purposes, therapeutic purposes, or both. Depending on the anatomic compartment, different procedures and specimen collection techniques are employed (Table 1.2) [2,5–15].

The collection of body fluid specimens can be performed on an outpatient basis or as part of an inpatient stay, which does require some patient preparation. To maintain patient safety and

Table 1.1 Advantages and Disadvantages of Different Body Fluid Test Ordering Schemes

	Advantages	Disadvantages
General	Single orderable with less administrative maintenance	Specimen source not clearly identified Interpretive comments may not be useful Laboratory may run risk of running specimen sources that are not validated
Specific	Clear order choices for analyte and body fluid source Interpretive comments tailored to specified source	Miscellaneous test ordering for fluid source not listed
Ordersets	Consolidates best practice orders for specified clinical procedures	May be limited by electronic order systems in use

minimize risks for complications, physicians and experienced medical professionals commonly employ the use of ultrasonography or other imaging techniques to help visualize the fluid compartment for fluid removal [8,10,11]. Collection devices such as needle and syringe combinations and thin, hollow needles and catheters are used. Fluids collected during the procedure are usually transferred to standard blood collection tubes, transport tubes, or liter bottles, depending on fluid burden. Collection of a matched serum or plasma sample is encouraged.

Before body fluid specimens are collected, the physician and assisting team should review required testing, appropriate container types, necessary volume, and consider any special collection requirements. For example, pleural fluid pH may be overestimated by 0.1–0.2 units if it is exposed to air; therefore, pleural fluids should be collected via syringe and immediately

Table 1.2 Body Fluid Collection Procedures and Fluid Types

Procedure	Mechanism of Fluid Removal	Recommended Fluid Designation	References
Paracentesis	Insertion of a needle through the abdominal wall into abdominal cavity.	Peritoneal	[5,6]
Thoracentesis	Insertion of a needle into the pleural space, between the surface of the lungs and interior chest wall.	Pleural[a]	[7]
Pericardiocentesis	Insertion of a needle into the pericardial space, the space between the visceral and parietal membrane layers of the heart.	Pericardial	[8,9]
Arthrocentesis	Insertion of a needle into the joint capsule or space.	Synovial[b]	[10]
Amniocentesis	Insertion of thin, hollow needle through the abdominal wall into the uterus.	Amniotic	[11]
Diagnostic Peritoneal Lavage	Insertion of a catheter into the peritoneal cavity to aspirate blood or fluid and subsequently infuse saline or fluid to lavage or wash the cavity for the rapid identification of intraperitoneal hemorrhage following penetrating trauma of the torso or abdomen.	Peritoneal Lavage	[12]

Lumbar Puncture, e.g., Spinal tap	Insertion of a needle into the subarachnoid space of the lumbar region of the spine.	Cerebrospinal	[13]
Aspiration	Insertion of a needle into a sac-like pocket of tissue, known as a cyst for diagnostic or therapeutic purposes.	Aspirate[c] Cyst Synovial[b]	[2]
Dialysis	Insertion of catheter through the abdominal wall into the peritoneal cavity to instill an osmotic dextrose solution for several hours prior to gravity drainage.	Peritoneal dialysate	[14]
Drain	Implanted or attached closed medical device for active removal of fluid from wound sites or near surgical sites.	Drain[c]	[2]
Washout following Fine Needle Aspiration	Needle washings collected after biopsy usually of lymph node or other tissue.	Fine needle aspirate biopsy (FNAB) washout[d]	[15]

[a] Designate right or left.
[b] Designate joint and right or left.
[c] Aspirate and drain are not fluid types; recommend designating a site or source.
[d] Recommend designating site or source of biopsy.

capped to preserve anaerobic conditions [16,17]. Depending on desired tests, the selection of appropriate blood collection or transport tubes is also important (Table 1.3). Standard blood collection tubes with gel barriers should not be used, because the functional separation of blood cell components is not necessary. Additionally, body fluid specimens collected in tubes with spray-coated anticoagulant should be inverted to ensure appropriate mixing.

Microbiologic studies show blood culture bottles—specifically aerobic and anaerobic bottle sets containing a proprietary medium (FAN) that removes a variety of microbial growth inhibitors from the specimen and enhances recovery of fastidious organisms—demonstrate superior performance with reduced time to detection compared with standard blood culture bottles and routine culture methods for bacteremia and fungemia in most body fluids [18–20]. Collectively, these studies demonstrate body fluids, like blood, should be directly inoculated into culture bottles, because doing so provides improved prognostic and diagnostic information for infectious disease management.

Ultimately, active dialog between healthcare professionals and the clinical laboratory, including the provision of resources such as an online test directory, a body fluid collection guide, or standardized electronic order collection instructions, may be useful to ensure successful body fluid collection and submission to the laboratory for testing (FAQ 1.1).

Labeling

As with any specimen collection, the quality of diagnostic results begins with proper patient identification. According to federal and/or state regulatory and clinical practice guidelines [21,22], laboratories should require that all specimens be labeled at the time of collection with at least two patient identifiers (PID):

1. PID:
    - a. Patient name (LAST NAME, FIRST NAME)
    - b. Unique number (e.g., medical record number)
2. Alternative PID:
    - a. Date of birth (MM/DD/YYYY)
    - b. Requisition number

Table 1.3 Common Standard Blood Collection Tubes Used for Body Fluid Collection

Tube Cap[a]	None	Clear[a]	Red	Lavender	Green
Tube Stopper[a]	Syringe	Marbled Red/ Light Gray[a]	Red	Lavender	Green
Possible Additive(s)	None Sodium heparin	None	Silicone coated	K2EDTA coated K3 EDTA coated K2EDTA liquid K3 EDTA liquid	Lithium heparin coated Sodium heparin coated Sodium heparin liquid
Testing	Pleural Fluid pH Synovial Fluid Microbiology	Chemistry	Chemistry	Hematology Crystal determination	Chemistry Crystal determination
Notes	Remove needle and cap		Silicone used as clot activator	Potassium EDTA used as an anticoagulant for blood cell determinations. Crystal determinations use liquid EDTA	Heparin used as an anticoagulant. Crystal determinations use liquid heparin.

[a] http://www.bd.com/vacutainer/labnotes/Volume18Number1/wall_chart.asp.
EDTA, ethylenediaminetetraacetic acid; *K*, potassium.

FAQ 1.1 How can the lab communicate collection instructions to body fluid collectors?

This is not easy! The solution may not be one size fits all and the strategy may be dependent on the care areas where collection takes place. Ultimately to achieve success, multiple strategies to communicate collection instructions may be required. From low-tech paper to high-tech electronic solutions, the following are some ideas to consider:

1. Sample Collection Handbook or Test Catalog—A resource that lists individual body fluid tests with the minimum and preferred volumes, sample container, and other applicable instructions. The pros include being rather quick to implement and with electronic deployment, there is no concern for out-of-date printed documents. The cons include not being specific to a procedure and putting the analysis and decision-making burden on the collector to decide which containers and volume of specimen to send for each procedure or collection.
2. Standard Operating Procedure or Quick Reference chart for collection area—A document that summarizes the common (and less common) tests collected by procedure or fluid type listing the minimum and preferred volumes, sample container, and other applicable instructions. The pros include being more of a quick reference guide that is customizable to the end user(s) for specified procedures. The cons include not easily accounting for dead space for tests that can be performed together in one tube. There may also be many different user groups to accommodate, and this could become unwieldy to manage and control.
3. Information System Collection Instructions—Build out the body fluid sample containers and system logic so that labels print that include container and test specific instructions for the collector to follow, much like blood draws. The pros include providing the most accurate just in time collection information for the collector. This is most critical for CSF collection. The cons include system limitations as well as effort to build out and maintain such systems.

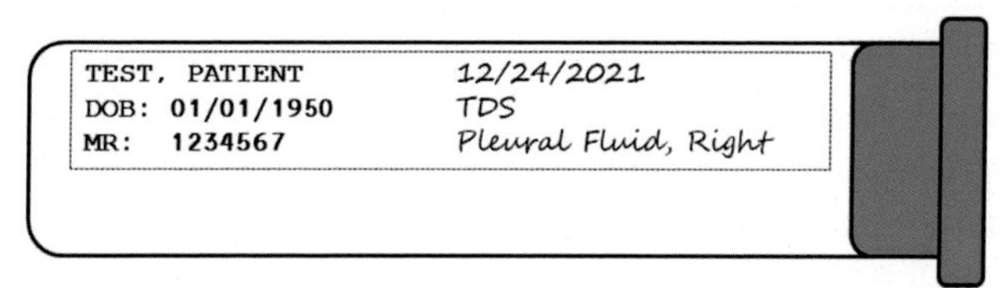

Fig. 1.3 Proper labeling of body fluid specimens. *DOB*, date of birth; *MR*, medical record.

Areas performing specimen collections should have access to barcode labels with this information. The practice of using preprinted labels should be discouraged because they increase the likelihood of patient misidentification, especially in shared procedural areas. It is good clinical and laboratory practice to label body fluid specimens with the date and time of collection, the initials of the collector, and the body site and fluid source (Fig. 1.3). Implementation of electronic ordering systems in procedural areas has increased patient safety as instrument-ready barcoded specimen labels can be affixed directly to specimen containers at the time of collection. Altogether, these items become increasingly important to ensure efficient processing, successful testing, and result reporting from the clinical laboratory.

Handling and Transport

Body fluid specimens should be handled with care and immediately delivered directly to the laboratory in a sealed secondary container (i.e., biohazard specimen bag). Specimen-receiving areas should review the specimens submitted to ensure appropriate labeling (Fig. 1.3) and forward fluid specimens immediately into testing areas to expedite analysis and reporting. In general, body fluid specimens are collected in containers acceptable for serum or plasma testing and transported in a manner consistent with those specimen types (FAQ 1.2) [23,24].

Body fluids collected using syringes and delivered to the laboratory for testing should be devoid of needles. One exception may be synovial fluid "dry taps," whereby the needle shaft may contain enough fluid for critical testing [25]. The use of syringes with lure-lock fittings to secure the cap and plunger locks or secondary

FAQ 1.2 What is the appropriate temperature to transport body fluid specimens to the laboratory?

In general, body fluid specimens can be transported to the laboratory at room temperature. A study investigating the stability of pleural fluid pH indicates that preserving the specimen on ice is unnecessary, as pleural fluid pH does not significantly change during the first hour after thoracentesis [23]. It has also been demonstrated that pleural fluid total protein, albumin, cholesterol, triglycerides, and glucose were stable up to at least 4 days when stored refrigerated (4°C) or at room temperature. Pleural fluid lactate dehydrogenase (LDH) showed no changes up to 2 days postcollection [24]. Ultimately, assessment of analyte stability during body fluid validation provides useful requirements for sample transport and storage. See Stability.

In addition, it may be worthwhile to compare hand-carried samples to those sent via pneumatic tube or compare an alternate sample container/additive. Pneumatic tube systems that transport specimens should be used with caution to ensure that these irretrievable specimens are not lost and results unaffected by this mode of transport (see Interferences).

means to protect the plunger from accidental engagement are necessary to ensure that body fluid specimens do not leak or are mistakenly expelled in transit. Ideally large volume specimens should be transferred to the necessary standard collection tubes or containers prior to delivery to the clinical laboratory, with the exception of cytologic examination. All containers should be labeled at the patient bedside before transporting to the laboratory for analysis (Table 1.4).

Specimen Assessment

Once the body fluid specimen arrives to the laboratory and the labeling, accessioning, and other administrative tasks are completed, the specimen should be assessed. Some attributes

Table 1.4 Body Fluid Collection Dos and Don'ts

Do	Don't
Deliver fluid specimens immediately to the laboratory in a secondary container (biohazard bag).	Send bulk fluid specimens for testing (bottles in liters). NOTE: This may be appropriate for Anatomic Pathology/Cytology, but not optimal for routine testing performed in the clinical laboratory.
Verify exogenous conditions (pneumatic tube system transport, addition of hyaluronidase) do not interfere with analyte recovery and reporting.	Deliver fluid specimens collected in syringe with exposed needles or capped needles.
Label specimen with two patient identifiers, date and time of collection, collector initials, and body fluid site and source.	
Refer to online test directory or available electronic orders to minimize preanalytical variables.	

to consider in the assessment include visual inspection for adequate volume, color, and clarity. It may be helpful to document the physical description or appearance and this information should be kept in the lab information system or attached to the results. To ensure all testing gets completed, aliquots may be necessary; as such, this is a good time to assess specimen viscosity. Fig. 1.4 demonstrates one option for assessing viscosity using a simple drop test, akin to the string test [26]. A "drop test" is a simple visual method to assess sample

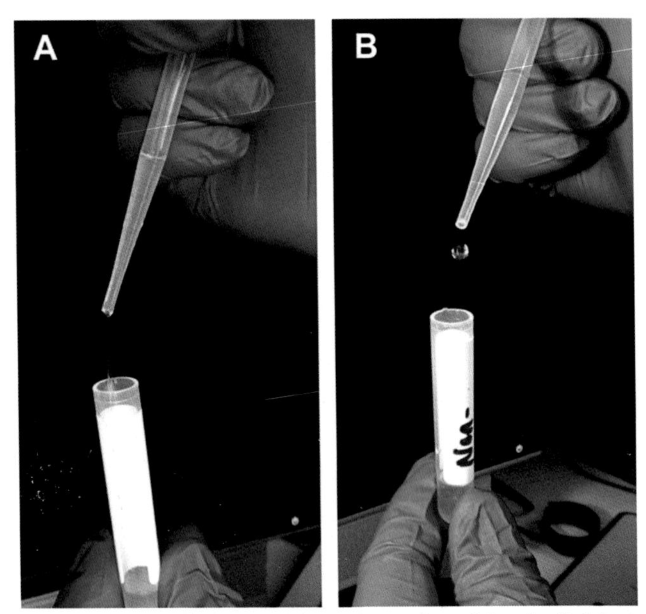

Fig. 1.4 Drop test for assessing viscosity of body fluids before placing on an automated instrument for analysis. (A) Viscous fluids have a longer string connecting to the pipette, whereas (B) nonviscous fluids have a uniform shape and no string. Reproduced with permission from Block DR, Florkowski CM. Body fluids. In: Rifai N, editor. Tietz textbook of clinical chemistry and molecular diagnostics. 6th ed. St. Louis, MO: Elsevier; 2018. p. 925–925.e35 [chapter 43], Copyright Elsevier (2017).

viscosity performed by drawing the body fluid up into a pipette and slowly squeezing drops of the fluid back into the tube [27]. If the drops are uniform in size and shape to that of serum/plasma, the sample viscosity is less likely to cause issue with the pipetting mechanism of the instrument. If the drops are not uniform in size and create a long string as it exits the pipette tip, further workup may be needed.

pH assessment is not commonly appreciated but laboratories may want to consider checking it at the time viscosity is assessed. Extreme pH ranges can be encountered in body fluids that are comprised, in part, of gastric contents, delayed transport, or from infectious sources demonstrating metabolic acidosis [28–30]. Enzyme activity can be impacted by extreme pH [31]. Screening body fluids or at least those with enzyme testing with pH paper to identify those with pH <7.0 should be considered. This ensures that body fluid results with decreased enzyme activity can be reported with either a comment or as the cancelation reason. For example, body fluid specimen check for pH measured <7.0, which decreases measured enzyme activity. In this case clinical correlation or comparison to enzymes in blood may be useful.

Specimen Pretreatment

Clinical laboratories commonly pretreat or perform upfront processing on body fluid specimens prior to analysis with the goal of producing a specimen that will give accurate reliable results. Some pretreatment examples to consider include centrifugation to remove turbidity, ultracentrifugation to remove lipemic interferences, filtering to remove particulate matter, or the addition of hyaluronidase to decrease sample viscosity (FAQ 1.3) [32,33].

The tube or aliquot intended for chemistry testing should be centrifuged. If there is particulate or cellular debris on the bottom, it is best practice to pour over or pipette the supernatant into a labeled, fresh aliquot tube. Ultracentrifugation of lipemic specimens may be appropriate, except when testing for chylous effusion (e.g., lipids) or when this pretreatment method interferes with recovery of analytes. It is important to validate whether preanalytic processing steps have the potential to impact results when testing body fluid specimens. These steps may be specific to the facilities, instruments, or other items used in your laboratory. See Interferences in Chapter 2 for guidance to validate specimen pretreatment practices to assess the impact on body fluid analyte measurements and result accuracy.

FAQ 1.3 What options are available for testing viscous samples?

Body fluids may be more viscous than other specimens received in the clinical laboratory. Hyaluronic acid is a lubricant molecule present on visceral surfaces to facilitate free movement in body cavities where body fluids are often collected (e.g., serous and synovial). Issues arise if specimens are too viscous to pipette accurately. There are three options for decreasing viscosity, including dilution, freeze/thaw, and hyaluronidase treatment [34]. Dilution is easy and possibly already part of the analytical validation; however, accurate pipetting is necessary so very viscous samples will present as a challenge. Gravimetric dilution is also discouraged as it could be misleading due to the differences in density of the sample and diluent. Freezing and thawing a viscous body fluid may be helpful to denature the proteins contributing to viscosity; however, it could be detrimental to the analyte of interest. Hyaluronidase is an enzyme that hydrolyzes hyaluronic acid. It can be added to viscous body fluids to decrease viscosity so that it may be pipetted more easily and accurately. Protocols include direct addition of powdered or liquid hyaluronidase, or addition of buffered hyaluronidase. These protocols have commonly been used for the pretreatment of synovial fluids for crystal examination and cell counting [32,33]; however, it is recommended to verify that the addition of hyaluronidase does not negatively affect analyte recovery as well as standardize a procedure for hyaluronidase pretreatment of any viscous body fluid specimen. Regardless, the hyaluronidase pretreatment procedure used should be incorporated into a standard operating procedure and be successfully reproduced by technologists.

References

[1] Petrides AK, Tanasijevic MJ, Goonan EM, Landman AB, Kantartjis M, Bates DW, et al. Top ten challenges when interfacing a laboratory information system to an electronic health record: experience at a large academic medical center. Int J Med Inform 2017;106:9–16.

[2] International Health Terminology Standards Development Organisation. SNOMED CT®, http://www.ihtsdo.org/snomed-ct/. [Accessed January 2022].
[3] Regenstrief Institute, Inc. LOINC., 2022, http://loinc.org/. [Accessed January 2022].
[4] Kee JL, Paulanka BJ, Polek CB. Handbook of fluid, electrolyte, and acid-base imbalances. 3rd ed. Clifton Park, NY: Delmar Cengage Learning; 2010.
[5] Runyon BA. Diagnostic and therapeutic abdominal paracentesis. In: Chopra SE, Robson KM, editors. UpToDate. Waltham, MA: Wolters Kluwer Health; 2022. http://www.uptodate.com/contents/diagnostic-and-therapeutic-abdominal-paracentesis. [Accessed January 2022].
[6] Runyon BA, Committee APG. Management of adult patients with ascites due to cirrhosis: an update. Hepatology 2009;49(6):2087–107.
[7] Daniels CE, Ryu JH. Improving the safety of thoracentesis. Curr Opin Pulm Med 2011;17(4):232–6.
[8] Jung HO. Pericardial effusion and pericardiocentesis: role of echocardiography. Korean Circ J 2012;42(11):725–34.
[9] Imazio M, Adler Y. Management of pericardial effusion. Eur Heart J 2013;34(16):1186–97.
[10] Punzi L, Oliviero F. Arthrocentesis and synovial fluid analysis in clinical practice: value of sonography in difficult cases. Ann N Y Acad Sci 2009;1154:152–8.
[11] Ghidini A. Diagnostic amniocentesis. In: Wilkins-Haug L, Levine D, editors. UpToDate. Waltham, MA: Wolters Kluwer Health; 2022. http://www.uptodate.com/contents/diagnostic-amniocentesis. [Accessed January 2022].
[12] Koyfman A, Long B. Peritoneal procedures. In: Roberts JR, Custalow CB, Thomsen TW, editors. Roberts and Hedges' clinical procedures in emergency medicine and acute care. 7th ed. Philadelphia, PA: Elsevier; 2019. p. 875–96.
[13] Wright BL, Lai JT, Sinclair AJ. Cerebrospinal fluid and lumbar puncture: a practical review. J Neurol 2012;259(8):1530–45.
[14] Rippe B. Peritoneal dialysis. In: Johnson RJ, Feehally J, Floege J, editors. Comprehensive clinical nephrology. 5th ed. Philadelphia: Saunders Elsevier; 2015. p. 1097–106.
[15] Suh YJ, Kim MJ, Kim J, Yoon JH, Moon HJ, Kim EK. Tumor markers in fine-needle aspiration washout for cervical

lymphadenopathy in patients with known malignancy: preliminary study. Am J Roentgenol 2011;197(4):W730–6.

[16] Venkatesh B, Boots RJ, Wallis SC. Accuracy of pleural fluid pH and PCO_2 measurement in a blood gas analyser. Analysis of bias and precision. Scand J Clin Lab Invest 1999;59(8):619–26.

[17] Sarodia BD, Goldstein LS, Laskowski DM, Mehta AC, Arroliga AC. Does pleural fluid pH change significantly at room temperature during the first hour following thoracentesis? Chest 2000;117(4):1043–8.

[18] Miller JM, Binnicker MJ, Campbell S, Carroll KC, Chapin KC, Gilligan PH, et al. A guide to utilization of the microbiology laboratory for diagnosis of infectious diseases: 2018 update by the Infectious Diseases Society of America and the American Society for Microbiology. Clin Infect Dis 2018;67(6):813–6.

[19] Mattei R, Savarino A, Fabbri M, Moneta S, Tortoli E. Use of the BacT/Alert MB mycobacterial blood culture system for detection of mycobacteria in sterile body fluids other than blood. J Clin Microbiol 2009;47(3):711–4.

[20] Simor AE, Scythes K, Meaney H, Louie M. Evaluation of the BacT/Alert microbial detection system with FAN aerobic and FAN anaerobic bottles for culturing normally sterile body fluids other than blood. Diagn Microbiol Infect Dis 2000;37(1):5–9.

[21] Clinical and Laboratory Standards Institute. Accuracy in patient and sample identification; approved guideline. 2nd ed. Wayne, PA: Clinical and Laboratory Standards Institute; 2019. CLSI document GP33.

[22] Centers for Medicare & Medicaid Services CMS, US Department of Health and Human Services. Clinical Laboratory Improvement Amendments of 1988 (CLIA). 42 CFR 493.1232—Standard: specimen identification and integrity. Washington, DC: CMS; 2003. https://www.govinfo.gov/content/pkg/CFR-2011-title42-vol5/pdf/CFR-2011-title42-vol5-sec493-1239.pdf. [Accessed January 2022].

[23] Rahman NM, Mishra EK, Davies HE, Davies RJ, Lee YC. Clinically important factors influencing the diagnostic measurement of pleural fluid pH and glucose. Am J Respir Crit Care Med 2008;178(5):483–90.

[24] Antonangelo L, Vargas FS, Acencio MM, Carnevale GG, Cora AP, Teixeira LR, et al. Pleural fluid: are temperature and storage time critical preanalytical error factors in biochemical analyses? Clin Chim Acta 2010;411(17–18):1275–8.
[25] Karcher DS, McPherson RA. Cerebrospinal, synovial, serous body fluids, and alternative specimens. In: Pincus MR, McPherson RA, editors. Henry's clinical diagnosis and management by laboratory methods. 24th ed. Philidelphia, PA: Elsevier Inc.; 2022. p. 510–38.
[26] Sanford SO. Arthrocentesis. In: Roberts JR, Custalow CB, Thomsen TW, editors. Roberts and Hedges' clinical procedures in emergency medicine. 7th ed. Philadelphia: Elsevier Saunders; 2019. p. 1105–24.
[27] Block DR, Florkowski CM. Body fluids. In: Rifai N, editor. Tietz textbook of clinical chemistry and molecular diagnostics. 6th ed. St. Louis, MO: Elsevier; 2018. p. 925-.e35.
[28] Holzer P. Taste receptors in the gastrointestinal tract. V. Acid sensing in the gastrointestinal tract. Am J Physiol Gastrointest Liver Physiol 2007;292(3):G699–705.
[29] Fura A, Harper TW, Zhang H, Fung L, Shyu WC. Shift in pH of biological fluids during storage and processing: effect on bioanalysis. J Pharm Biomed Anal 2003;32(3):513–22.
[30] Mishra EK, Rahman NM. Factors influencing the measurement of pleural fluid pH. Curr Opin Pulm Med 2009;15(4):353–7.
[31] Nandakumar V, Dolan CT, Baumann NA, Block DR. Effect of pH on the quantification of common chemistry analytes in body fluid specimens using the Roche cobas analyzer for clinical diagnostic testing. Am J Clin Pathol 2021;156(5):722–7.
[32] Brunzel NA. Appendix C. Body fluid diluent and pretreatment solutions. In: Fundamentals of urine & body fluid analysis. 3rd ed. St. Louis: Elsevier Saunders; 2013. p. 422–3.
[33] Clinical and Laboratory Standards Institute. Body fluid analysis for cellular composition; approved guideline. CLSI document H56-A, Wayne, PA: Clinical and Laboratory Standards Institute; 2006.
[34] Clinical and Laboratory Standards Institute. Analysis of body fluids in clinical chemistry; approved guideline. CLSI document C49-B, Wayne, PA: Clinical and Laboratory Standards Institute; 2018.

Analytic Validation

If you are reading this *Quick Guide*, you probably acknowledge that body fluid tests require validation but may still question if it is really necessary. With common exceptions being urine and cerebrospinal fluid, manufacturers of in vitro diagnostics (IVD) do not routinely specify the performance characteristics of nonblood (i.e., specimens other than serum or plasma) in the instrument and/or assay instructions for use. Testing body fluids is off-label and is considered a modification of an FDA-approved test system (Fig. 2.1); therefore, to ensure that accurate results are produced so that providers are best able to manage patients safely and effectively, laboratorians should validate assay performance according to the modifications for fluid types in which they intend to provide results (FAQ 2.1).

Based on the Clinical Laboratory Improvement Act of 1988, the College of American Pathologists (CAP) expects that studies establishing accuracy/trueness, precision, analytical sensitivity and specificity, reportable range, reference range, and interferences should be performed [1,2]. Other laboratory accreditation agencies (The Joint Commission, Commission on Office Laboratory Accreditation (COLA), International Organization for Standardization (ISO)) have similar requirements; however, the decision to validate and implement any modified method should be made by the laboratory director. Furthermore, the CLSI guideline for the Analysis of Body Fluids in Clinical Chemistry suggests that the validation plan developed by the laboratory director should include a risk-based approach which takes into consideration how the testing is used clinically and what is practical given laboratory resources [3]. In support of these regulatory and accreditation guidelines, the following chapter outlines effective analytical validation studies and details data analysis strategies for improved quality in the laboratory.

Accuracy/Trueness

The recommended experiments outlined in this *Quick Guide* assume that the body fluid assay being validated already meets the manufacturer's specifications of performance for the intended

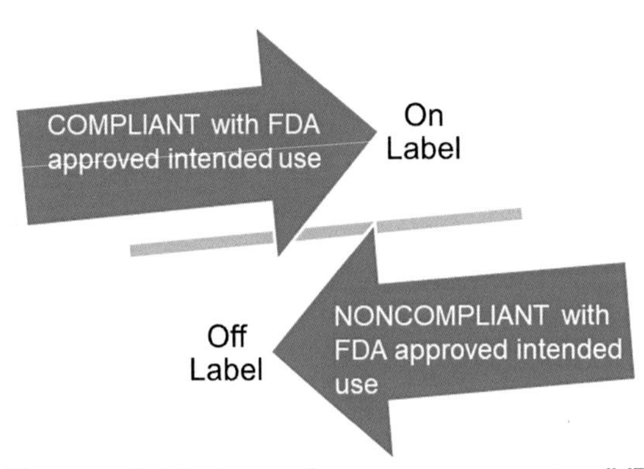

Fig. 2.1 On vs off-label use for assay reagents. *IVD*, in vitro diagnostics.

FAQ 2.1 If serous = serum, then serous body fluids don't really need to be validated, right?

Admittedly, this can be confusing; however, there is regulatory and accreditation guidance on the topic (COM.40620) which requires discernment [1]. There is a tendency to get hung up on the statement that suggests the "method performance specifications" for the approved matrix (blood, serum, urine, etc.) can be used for body fluids if the lab demonstrates the absence of matrix interferences. Since it is a reasonable assumption that serous body fluids, derived from ultrafiltrates of plasma, have many similar characteristics this "should" be true. However, it is important to understand that a method performance specification is in fact a performance claim. For example, if the assay package insert claims its imprecision in serum at 3 g/dL is CV = 5%, we would expect that by modifying the assay's intended use, body fluids of similar concentration should at least meet this performance. The bottom line is you must do the study to make that claim, so please do not fall victim to this fallacy.

sample type and ideally is also being used to report clinical results. Thus assessment of accuracy or trueness involves confirming that an analyte can be accurately measured in an alternate matrix. These experiments are important to perform thoroughly and best to conduct early in the evaluation because these studies lay the groundwork for the rest of the analytic validation. Method validation for accuracy should be conducted utilizing a variety of body fluid types and sources, reflecting the most frequent and clinically relevant fluid types (FAQ 2.2) [4–6].

The predominant issue contributing to potential inaccuracy and recovery of the analyte of interest is the presumed heterogeneous nature and composition of the body fluid matrix itself. Inaccuracy due to matrix composition may be multifactorial and includes variables such as pH [8], ionic strength, protein, and lipid content—all of which can influence the solubility of the analyte of interest or potentially the rate of enzymatic reactions. In addition, body fluid specimens are often viscous, which may

FAQ 2.2 Which body fluid types and tests should be validated?

The answer is analyte dependent, and it is important to realize not every analyte has clinical utility beyond its measurement in serum or plasma. The laboratory is encouraged to use some discretion when determining which body fluid types to validate and literature references should be consulted to determine best practice. Reviews have been published outlining the most useful analytes to measure in body fluids [4–6]. A more recent review critically evaluated the methods and study designs for analytes measured in serous fluids and may be useful when considering body fluid test validations [7]. When determining utility of nonserous body fluid tests, a similar critical approach is recommended. Here is a list of important things to consider:

1. Is this test clinically indicated?
2. Are alternate validated tests available with equal value?
3. Will the results of the test inform or change how the patient is managed?
4. What is the required turnaround time (TAT) for appropriate patient management?
5. Can this be referred to a reference laboratory?

FAQ 2.3 Should I worry about body fluids clogging up the instruments and/or potentially producing inaccurate results?

This is a valid concern and steps should be taken to minimize the potential as much as possible. Body fluids should be centrifuged, aliquoted, visually inspected, and assessed for increased sample viscosity. See Chapter 1 for a thorough discussion of these preanalytic processing steps. Essentially, it is critical to take these steps and deal with sample viscosity concerns (see FAQ 1.3) adequately prior to introducing the sample to an automated analyzer which should alleviate much of the risk to the instrument and accuracy of results.

result in sampling errors. In this situation, results may not be reproducible upon repeat testing (FAQ 2.3).

The recommended approach to validating accuracy/trueness of body fluid assays de novo, where a previous validated method does not exist, includes performing at least one if not all of the following recovery experiments. Recovery studies would not be necessary if you are validating a new instrument to replace one that has already demonstrated it meets the method performance specifications (see Method Comparison). This also assumes there is a decent level of confidence for how thorough prior studies were conducted and documented.

Spiked Recovery

A series of body fluid specimens are spiked with a known concentration of analyte to determine if the body fluid matrix affects the recovery of the measured analyte.

Specimen preparation:

1. Select at least three specimens with a low concentration of the analyte of interest.
2. Spike specimens with material with high concentration of analyte. Spiking is ideally done using commutable material (e.g., behaves and possesses properties akin to clinical specimens) such as serum, but when difficult to accommodate can be

done with calibrators, quality control (QC) materials, or other standard material of known concentration.
3. Ensure that the volume change upon addition of the spiking material is less than 5% to minimize the change in matrix composition [3,9].
4. Create a series of spiked samples with increasing analyte concentration within the measurable range of the assay and mix well (Fig. 2.2).

NOTE: *Include and prepare control sample(s) using a diluent to account for any dilutional effects of spiking.*

Test:
1. Measure the analyte of interest in prepared samples. Consider measuring in at least duplicate and averaging results.
2. Repeat this process for each body fluid type/source and analyte to be validated.

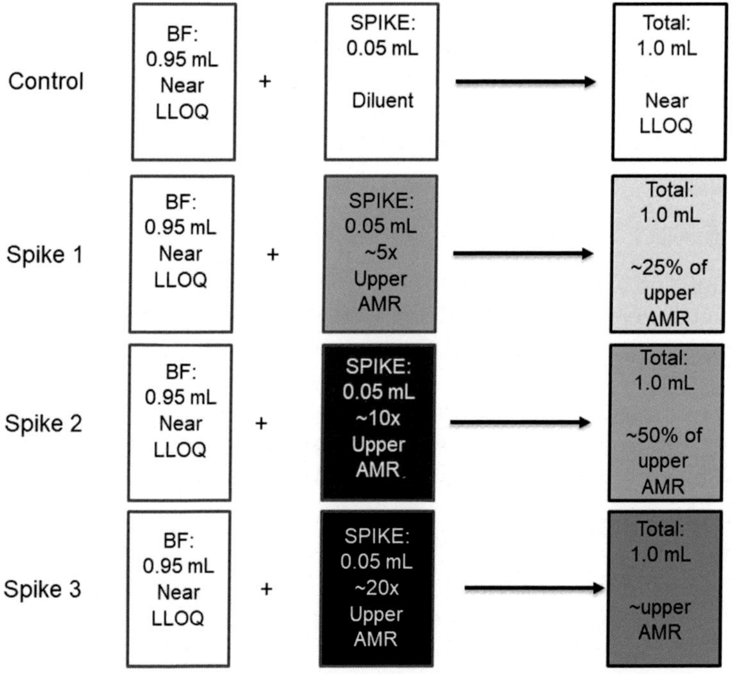

Fig. 2.2 Scheme for preparing spiked recovery samples. *AMR*, assay measurement range; *LLOQ*, lower limit of quantitation.

Table 2.1 Example Data for Spiked Recovery Study

Sample	Final (Units)	Initial (Units)	Added (Units)	% Recovery
1	59	4	50	110
2	79	3	75	101
3	98	1	100	97

Note: It is suggested to use an initial concentration that is low yet within the measurement range (e.g., within 100% of the lower limit of quantitation).

Analyze:

1. Calculate % recovery = [(final-initial)/added] × 100% (Table 2.1), where:
   a. *Final:* Final measured concentration after spiking.
   b. *Initial:* Initial measured concentration in the prepared control sample spiked with diluent.
   c. *Added:* Concentration of analyte used to spike into sample; it is imperative this concentration is known or can be measured.

Mixed Recovery

High and low concentration specimens are mixed in equal proportions to create a series of samples whose analyte concentration is equally spaced across the measurement range of the assay. This study design is advantageous because the analyte recovery is assessed in the true fluid matrix state. Additionally, this study serves to confirm a linear response measuring the analyte in body fluids (see Linearity/Reportable Range).

Specimen preparation:

1. Select high- and low-concentration specimens whose concentrations are within the assay measurement range, ideally near the intended upper and lower limits.
2. Mix specimens in 1:1 ratios to create a sample series whose concentrations are equally spaced over the measurement range and mix well (Fig. 2.3).

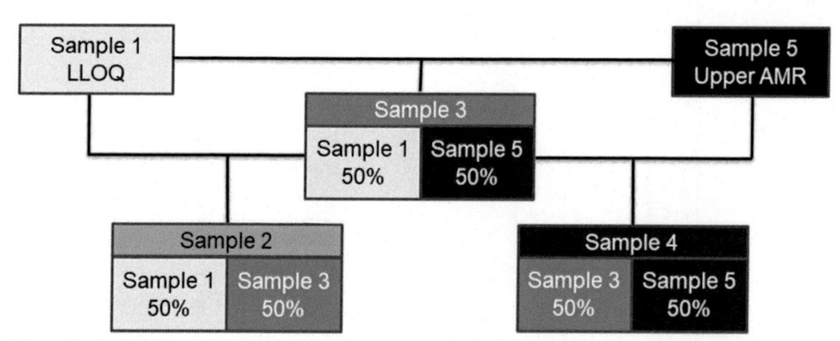

Fig. 2.3 Scheme for preparing mixed recovery samples. *AMR*, assay measurement range; *LLOQ*, lower limit of quantitation.

Test:

1. Measure the analyte of interest in the sample series. Consider measuring in at least duplicate and averaging results.
2. Repeat this process for each body fluid type/source and analyte to be validated.

Analyze:

1. Calculate the expected analyte concentration for each sample, setting the expected concentration equal to the measured concentration in one sample (i.e., 100% recovery) (Table 2.2). ***NOTE:*** *Refrain from setting both end points equal to the measured concentration. The advantage of selecting the middle point is being able to assess over/underrecovery at the upper and lower limits of the measurement range.*
2. Calculate percent recovery = (measured/expected) × 100% (Table 2.2).

Dilution Recovery

A high specimen whose concentration is within the measurement range is serially diluted with diluent to determine the recovery of the analyte. This study confirms linear response of measured analyte and helps to confirm reportable range (see Linearity/Reportable Range).

Specimen preparation:

1. Select an appropriate diluent for the assay. Options to consider include those recommended by the manufacturer for the intended specimen type or another suitable diluent.

Table 2.2 Example Data for Mixed Recovery Study

	[Measured] (Units)	Delta (Δ)		[Expected] (Units)		% Recovery
	Data	Formula	Data	Formula	Data	
A	2	$B-A$	25	$C-(2\times\Delta ave)$	1.0	200
B	27	$C-B$	21	$C-\Delta ave$	24.5	110
C	48	n/a	n/a	C	48	n/a
D	71	$D-C$	23	$C+\Delta ave$	71.5	99
E	96	$E-D$	25	$C+(2\times\Delta ave)$	95.0	101
		Δave[a]	23.5			

[a] Average delta (Δave) is the average analyte concentration difference between equally spaced samples.
NA, not applicable; [], concentration.

A protein-based diluent such as 7% bovine serum albumin solution may be ideal, especially for assays with large dynamic ranges (enzymes) that may exhibit better recovery upon large dilution. Preservation of the protein-based matrix simulates serum/plasma and may support more efficient recovery of intended analyte in body fluids.

2. Dilute a specimen at the upper limit of the measurement range serially (1:2, 1:4, etc. or 1:10, 1:100, etc. as appropriate for the range of the assay) and mix well (Fig. 2.4).

Test:

1. Measure the analyte of interest in the sample series. Consider measuring in at least duplicate and averaging results.
2. Repeat this process for each body fluid type/source and analyte to be validated.

Analyze:

1. Set the undiluted concentration (neat sample) equal to the expected concentration to calculate the theoretical concentrations of analyte in each diluted sample (Table 2.3).
2. Calculate percent recovery = (measured/expected) × 100%.

Recovery Study Summary

Plot:

1. Plot recovery for all accuracy experiments as a function of measured analyte concentration.
2. Determine where in the measurement range the recovery appears to be poor (Fig. 2.5).

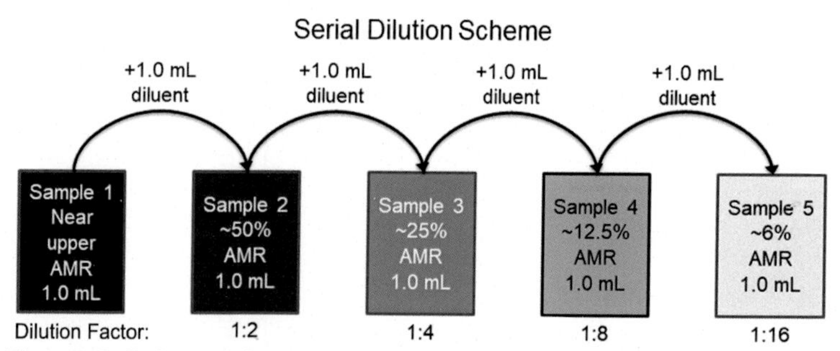

Fig. 2.4 Scheme for preparing serially diluted samples. *AMR*, assay measurement range.

Table 2.3 Example Data for Dilution Recovery Study

Sample	Dilution Factor	Measured Concentration (Units)	Expected Concentration (Units)	Recovery (%)
1	n/a	101	101	n/a
2	1:2	50	50.5	99
3	1:4	23	25.25	91
4	1:8	11	12.6	87
5	1:16	4	6.3	63

n/a, not applicable.

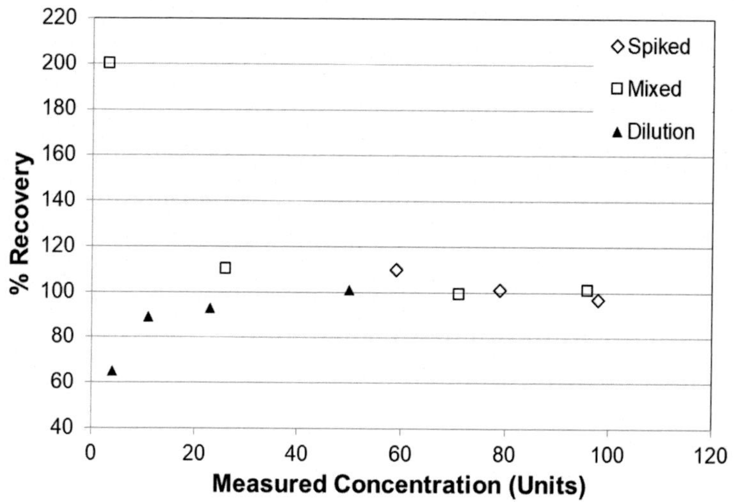

Fig. 2.5 Analysis of the percent recovery from accuracy experiments. Poor recovery is observed at a low concentration for both mixing and dilution, suggesting the measurement range should be ~10 rather than 1 (see Linearity/Reportable Range).

Acceptance criteria:

1. Criteria should be determined by the laboratory director.
2. It may be based on expected performance in an acceptable matrix (plasma, serum, urine) either from the manufacturer's claims, control experiments performed on serum/plasma/urine, or available peer-reviewed literature (see FAQ 2.2).
3. In cases in which the acceptance criteria are not met, the laboratory director should assess the impact of bias and poor recovery on the clinical decision limits and the interpretation of results in the body fluid matrix being studied.

Method Comparison

Method comparison to a reference method is a straightforward study to verify accuracy/trueness. However, body fluid tests do not have true reference methods and as such should not be relied on in de novo body fluid validations. A previously validated method could serve as a suitable option, such as when replacing instruments. Laboratories replacing an instrument validated for body fluid testing should first ensure prior validation studies were conducted and documented well. If any gaps exist or documentation missing, consider conducting studies to ensure method validation meets claimed performance. Second, ensure that the assay intended to be used for body fluid testing performs acceptably for its intended use (e.g., serum, plasma, urine). This will be natural if both on-label sample types and body fluid testing will be conducted on the same instrument(s). If, however, the laboratory intends to isolate body fluid testing, then they should verify the assay performance is acceptable in approved specimen types in accordance with manufacturer's instructions for use in addition to conducting the body fluid comparison and validation studies.

Specimen preparation:

1. Select at least 55 samples for the method comparison study. A minimum of 40 samples should be the approved sample type (serum, plasma, urine) and a minimum of 15 body fluid samples, ideally spanning the measurement range. The body fluids selected will be influenced by the analyte and sources to be tested (see Table 2.4 for guidance).

Table 2.4 Suggested Design Guide for Body Fluid Method Comparison Studies

Analyte	Specimen Type/Source[a]
Albumin	Peritoneal > Pleural > Drain
Amylase/Lipase	Peritoneal > Drain > Pleural > Pericardial
Bilirubin, Total	Peritoneal > Drain > Pleural > Pericardial
Cholesterol/ Triglycerides	Peritoneal, Pleural > Drain > Pericardial
Creatinine/Urea Nitrogen	Drain, Peritoneal, Peritoneal Dialysate > Pleural > Pericardial
Glucose	Peritoneal, Peritoneal Dialysate, Pleural > Drain, Amniotic > Pericardial, Synovial
LDH/Total protein	Pleural > Peritoneal, Drain > Pericardial, Synovial

[a] Listed from most to least common and clinically significant fluid sources for validation.

Test:

1. Measure the analyte(s) of interest in the sample series.

Plot:

1. Reference method (x) vs new method (y) (Fig. 2.6A) NOTE: Serum and body fluid results can be plotted together and/or separately.
2. Plot bias from different sample types using different symbols to ensure there are no consistent outliers (Fig. 2.6B).

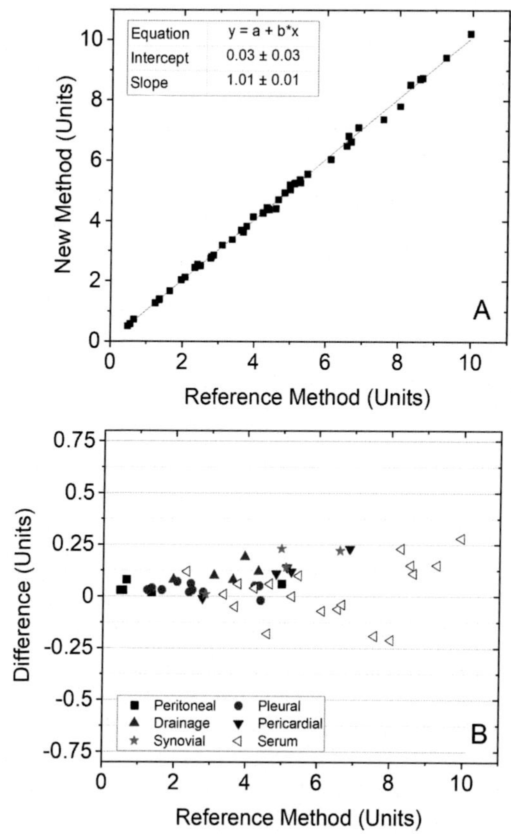

Fig. 2.6 Regression (A) and bias plot (B) analysis when conducting method comparison studies for instrument replacement.

Analyze:

1. Perform regression analysis to determine slope, *y*-intercept, and correlation coefficient.
2. Calculate difference and % difference between the reference method and the new method for each individual sample.

Acceptance criteria:

1. Criteria should be determined by the laboratory director. Options include:
   a. Expected analytical performance in an accepted matrix (e.g., plasma, serum, urine).
   b. Bias within allowable limits for the body fluid matrix being studied wherein the interpretation is not impacted. A majority (e.g., >90%) of samples tested should be within these limits.

Analysis and interpretation of the method comparison should be straightforward if replacing with same vendor equipment; however, if faced with changing vendors, then understanding bias between methods will be critical. There may be observed differences between methods which can be attributed to calibration traceability and other differences in methods including measurement range limitations that laboratories need to appreciate [10].

Other Considerations: Methods for Assessing Accuracy/Trueness

A traditional method comparison study to another laboratory can have limited value when used to establish accuracy/trueness for body fluid validation. The limitation lies in the available information to ascertain the performance characteristics demonstrated in body fluids at the comparator lab. For example, both methods will likely provide similar results and demonstrate similar bias, should one exist, and it is for this reason, laboratories should perform their own accuracy/trueness evaluation. One exception to comparing with another laboratory may be in a hospital network where testing (analytical systems and reagents) is standardized. If one site demonstrates acceptable performance for body fluid validation studies, they could provide reference body fluid specimens with assigned values to satellite sites to assist with required accuracy/trueness studies.

FAQ 2.4 How should we handle requests for body fluid types whose source we didn't validate?

First, body fluids should have an identifiable source specified. If it is missing or too general (e.g., "body fluid") it is prudent to contact the provider for clarification as needed. For body fluid test requests where the source is listed, yet not specifically validated (examples include blister fluid, perihepatic fluid, etc.), you may still consider contacting the provider to determine whether the test has utility and how results will be used. If the desire is to report results, then a suitable send-out test could be identified. If that search reveals no options, then laboratories may want to consider performing serial dilution or spiking recovery studies to verify results are accurate. In these cases, the laboratory director should review results and provide any necessary interpretation (see Chapter 3—Reporting Results) and laboratory body fluid testing policy should reflect this exception in practice.

Using proficiency testing (PT) material (unless it is an accuracy-based assessment) to establish accuracy/trueness claims should be avoided. PT is fraught with similar challenges demonstrating comparability within a peer group and has noncommutability concerns. Although participation in a PT program is required for ongoing quality assurance (see Chapter 3), performance evaluations that include PT material should not replace more robust experiments of accuracy.

Lastly, it is important to recognize accuracy evaluations are performed and documented to ensure clarity of acceptable body fluid sites and sources. See FAQ 2.4 for recommendations when this isn't the case.

Linearity/Reportable Range

The assay measurement range is the concentration range over which an analyte can be measured accurately. This typically occurs where the assay demonstrates a linear signal response to the analyte concentration. Depending on the capabilities of your instrument or laboratory information systems, it is often easiest for body fluid assays to have the same measurement range as demonstrated for serum/plasma; however, this is not required. If the measurement range is the same or narrower than serum/plasma, then it is sufficient to use the mixing and dilution data generated in the accuracy experiments to assess linearity and verify the assay measurement and clinical reportable range.

Analyze:

1. Plot measured (*x*) vs expected (*y*) analyte concentration from both mixing (Fig. 2.7A) and dilution (Fig. 2.7B) recovery experiments.
2. Perform regression analysis to determine the slope, intercept, and correlation coefficient.

If the measurement range is intended to be wider than the current range used for serum/plasma, then it is advisable to perform validation with manufactured linearity material with assigned concentration values for serum to verify the assay is accurate and linear throughout the intended measurement range. Conducting the mixing and dilution studies with body fluid specimens ensures the performance claim for a body fluid matrix has been evaluated and is acceptable and assures a lack of matrix interference.

The reportable range is established by determining the concentration that can be accurately reported for specimens requiring dilution whose undiluted analyte concentration is above the linear range. The need for and extent of dilutions required for a body

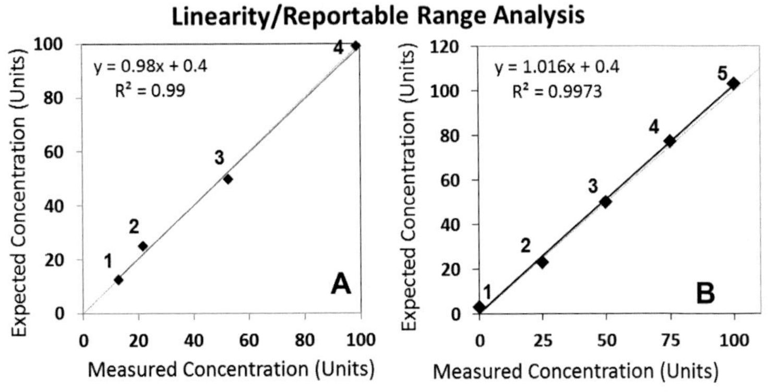

Fig. 2.7 Linear regression analysis to determine reportable range from (A) mixed and (B) serially diluted samples.

fluid assay may or may not be similar to the dilutions needed with serum/plasma. This need will depend on the clinical interpretation of results. The maximum dilution is determined by selecting a high concentration specimen near—but not outside—the measurement range and performing serial dilutions. Doing so ensures that the undiluted concentration can be measured with certainty and, as such, should be used to calculate the expected concentrations of analyte in diluted samples. In such cases, a "greater than" result may be reported if clinically indicated and further dilutions are unnecessary. If automated instrument or pipetting systems perform the dilutions, then the accuracy of those dilutions is best compared with results obtained using manual dilution.

Other Considerations: Methods for Assessing Linearity/Reportable Range

In clinical practice, further dilutions may be requested and performed on a case-by-case basis. The goal is to demonstrate at least three results are within the measurement range and that the response is linear. If exceptions for reporting are made, approved practice deviations should be included in standard laboratory operating procedures. Consider incorporating laboratory director review of such results prior to reporting or use predetermined acceptance criteria in a standard operating procedure.

Precision

Conducting precision experiments helps to ensure that the assay demonstrates acceptable reproducibility within a given matrix. Inter- and intra-assay precision experiments can be performed with a protocol like that used for plasma or serum precision studies (FAQ 2.5).

Laboratories may use CLSI EP15-A3 [11] to verify precision performance against manufacturer's claims and EP Evaluator or similar statistical software to simplify the calculation of repeatability (intra-assay) and within-laboratory (interassay) precision. ***NOTE****: Using CLSI EP5-A3 for establishing acceptable precision performance may be overkill unless the assay is being solely used for body fluid testing or where manufacturer's claims are not available* [12]. The exact protocol is up to the discretion of the laboratory director. Depending on computing resources, one of the

FAQ 2.5 Can body fluids of similar origin be pooled together to get sufficient volume for validation?

Ideally, the answer is no, because of the potential to dilute interferences and/or differences in fluid matrix (serous vs non-serous). However, in the real world, labs need to do the best with the resources available and pooling similar fluid types together is one way to do this. It is recommended to minimize the number of samples pooled and experiments where pooled samples are used as much as possible. In such cases where large volumes of fluids may be needed laboratories may want to consider reaching out to anatomic pathology areas where large bottles of fluids may be submitted for cytology evaluations.

following options is suggested. One or more runs per day, in at least duplicate, for at least 10 days. One run per day, in singlicate, for at least 20 days and a single run with 20 replicates. Prior to beginning precision experiments, it is important to know analyte stability because the protocol involves making and storing aliquots to be run over multiple days (see Stability).

Specimen preparation:

1. Choose a minimum of two different body fluid types (each ideally from the same patient/collection). Ideally, select ones that have:
   a. sufficient volume to prepare 10 aliquots with enough volume to run multiple times (typically 5–10 mL, dependent on dead volume).
   b. analyte concentrations that are low, high, and/or near the medical decision point (MDP).
2. Aliquot each fluid into at least 10 aliquots.
3. Store them under conditions where the analyte is most stable (i.e., refrigerated, frozen).

Test:

1. Remove an aliquot from storage.
2. Adhere to a consistent protocol that can be operationalized for clinical testing, to thaw (if indicated), mix, and centrifuge the sample prior to testing.

3. Measure the analyte according to the desired protocol.
4. Repeat this process for each body fluid type/source.

Analyze:

1. Calculate repeatability and within-laboratory precision using statistical software or calculate standard deviation and coefficient of variation for within-run and between-run precision.

Acceptance criteria:

1. The laboratory director should determine these criteria.
2. Acceptance criteria for precision studies may be compared with the manufacturer's performance claims for the assay in serum/plasma or QC.
3. Consider that the decision limits and interpretation may be different for body fluids than for serum concentrations, so it may be necessary to prepare a QC pool that targets a particular region of the measurement range in addition to QC run(s) during normal operations.

Analytic Sensitivity

The determination of precision at the lower limit of quantitation (LLOQ) helps define the lowest concentration of an analyte that can be reliably measured and thereby reported. Poor low-end precision may significantly impact measurement range decisions at low concentrations for analytes such as albumin and total protein. However, it may not affect the interpretation of some body fluid assays where extreme elevations in an analyte are expected in disease states (e.g., pancreatic enzymes). Conducting LLOQ experiments offers some advantages. Such experiments confirm the assay has similar precision in a body fluid matrix compared with serum or plasma and that the same lower reportable range may be used for both applications. It is advisable to conduct whichever studies your accrediting body deems necessary.

Specimen preparation:

1. Choose a low concentration specimen near the LLOQ.
2. Aliquot the body fluid into at least five tubes with sufficient volume to measure in quadruplicate.
3. Store aliquots under conditions where the analyte is most stable (i.e., refrigerated, frozen).

Test:

1. Each day for five consecutive days, remove an aliquot from storage and measure the analyte in quadruplicate.
2. Repeat this process for each body fluid type/source and analyte to be validated.

Analyze:

1. Calculate mean, standard deviation, and coefficient of variation.

Acceptance criteria:

1. The laboratory director should determine criteria.
2. For most analytes, the goal is a coefficient of variation below 20%; however, this value may vary depending on the application. ***NOTE:*** *Consider comparing imprecision in body fluid with serum/plasma if there are concerns.*

Interferences

Analytical specificity evaluates the impact of endogenous agents and exogenous variables on the accuracy of the assay. Examples of endogenous interferences include hemoglobin, bilirubin, and lipids. Examples of exogenous interferences and variables include preanalytic steps taken to decrease sample viscosity or turbidity (see Specimen Pretreatment) or even pneumatic tube transport (see Handling and Transport). The threshold of tolerance for any interference can be determined by assessing the impact of clinical interpretation at varying concentrations of the interfering agent or variable condition.

It is important to consider whether additional endogenous interfering substances are present beyond hemoglobin, bilirubin, and lipids that may be body fluid specific, such as meconium contamination in amniotic fluid. The method for determining the concentration of interfering substance is also important to consider. Automated instruments measuring hemoglobin, bilirubin, and lipemic interference for serum/plasma assays may be the preferred method for measurement in body fluids. Interpretation and MDPs may be different in a body fluid compared with serum; therefore, it is important to determine the appropriate cutoffs for these interferences in body fluids and not simply adopt the serum limits or use package insert claims without further validation. The laboratory must also decide when and how to report results for specimens not meeting these criteria with standard operating procedures clearly outlining the entire process. Considering body fluids are usually irretrievable specimens and subsequent recollection may not yield the same output or require repeated invasive procedures, laboratories should consider if the test should be canceled on a hemolyzed body fluid or if the result should be reported with a comment stating the effect that the interference may have on the results.

Specimen preparation:

Endogenous

1. Select body fluid specimens that contain a low and high concentration of the analyte to be tested for interference, a concentration at the MDPs, or both. The concentration and number of specimens to test are dependent on the degree of interference caused by spiking or the variable condition. For example, hemolysis interference will influence lactate dehydrogenase (LDH) activity to a much greater extent than creatinine concentration.
2. Endogenous interferent solution preparation
   a. Hemoglobin for detecting the effect of hemolysis
      - Prepare a hemolysate solution by lysing washed red packed cells. Dilute with saline or approved diluent to achieve the desired stock concentration needed for preparing the max spike sample (Fig. 2.8).

   b. Bilirubin for detecting the effect of icterus
      - Prepare a stock bilirubin solution using purchased conjugated and unconjugated bilirubin or high patient specimens. Dilute with saline or approved diluent to achieve the desired stock concentration needed for preparing the max spike sample (Fig. 2.8).

   c. Lipids for detecting the effect of lipemia
      - Prepare a stock solution of commercially available lipid solution or high patient specimens. Dilute with saline or approved diluent to achieve the desired stock concentration needed for preparing the max spike sample (Fig. 2.8).

   Note: The Spike Max should contain interferent concentration about 20-fold greater than the targeted threshold to account for dilution into the body fluid.

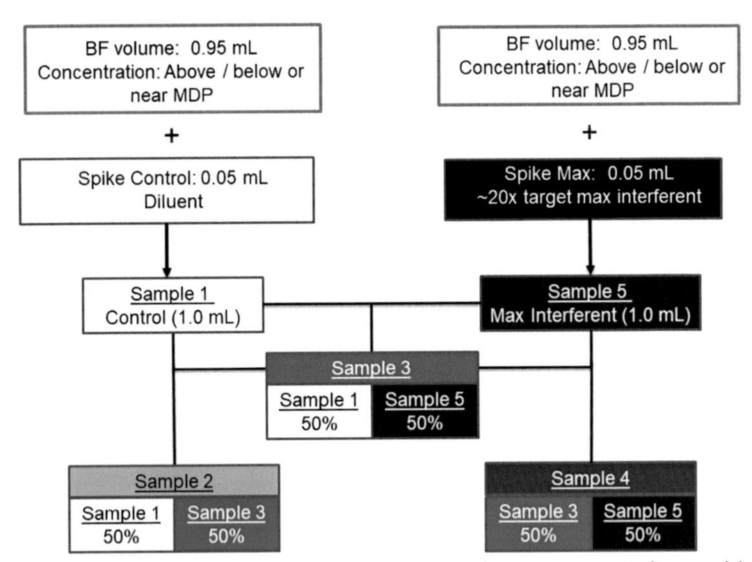

Fig. 2.8 Scheme for preparing interference samples with increasing concentration of endogenous substances, including hemoglobin, bilirubin, or lipids. *MDP*, medical decision point.

3. Split a single body fluid sample into equal volume portions (see gray outlined boxes in Fig. 2.8).
4. Prepare the maximum interference spiked sample (Sample 5) by adding a small volume (<5%) from the interferent solution to create a sample with interferent concentration at least 20% above the intended limit [3,9].
5. Prepare the lowest interference sample (Sample 1) which will serve as the control sample that accounts for analyte dilution from interferent spiking.
6. Mix Sample 5 with Sample 1 in equal ratios to prepare a sample set that has an expected analyte concentration represented in the control and increasing interferent concentration (Samples 2, 3, and 4) and mix well (Fig. 2.8).

Exogenous

1. Define the variable condition(s) to test (e.g., before and after addition of hyaluronidase or transport by hand vs pneumatic tube transport to the laboratory).

2. Select at least 3 body fluid specimens that contain analyte concentration near the MDP.
3. Split each body fluid specimen to be tested into two aliquots.
4. Apply the variable condition (test) to one aliquot and leave the second aliquot unaltered (control).

Test:

Endogenous

1. Measure the analyte concentration in each sample. Consider measurement in duplicate and averaging the results.
2. Measure or estimate the interferent concentration in each sample.
3. Repeat this process for each interferent and body fluid concentration.

NOTE: *A different body fluid type can be used to span the measurement range of the assay* (e.g., *peritoneal fluid with a low protein concentration and pleural fluid with a higher protein concentration).*

Exogenous

1. Measure the analyte concentration in the test and control aliquots. Consider measurement in duplicate and averaging the results.
2. Repeat this process for each variable and body fluid concentration.

Analyze:

Endogenous

1. Calculate difference or % difference = (spiked − control)/control × 100% (Table 2.5).

Exogenous

1. Calculate difference or % difference = (variable condition applied − control condition)/control condition × 100%.

Plot:

1. Plot difference or % difference vs interferent concentration for the sample series.
2. Identify the interferent tolerance limit where the result is outside acceptable limits or changes the interpretation (normal to abnormal or vice versa; Fig. 2.9).

Table 2.5 Example Data for Interference Spiking Study

Sample	Interferent Concentration (Units)	Analyte Concentration (Units)	Difference	% Difference (×100%)
1/Control	0	100	0	0
2	250	105	5	105–100/100 = 5
3	500	111	11	111–100/100 = 11
4	750	119	19	119–100/100 = 19
5/Max Interferent	1000	122	22	122–100/100 = 22

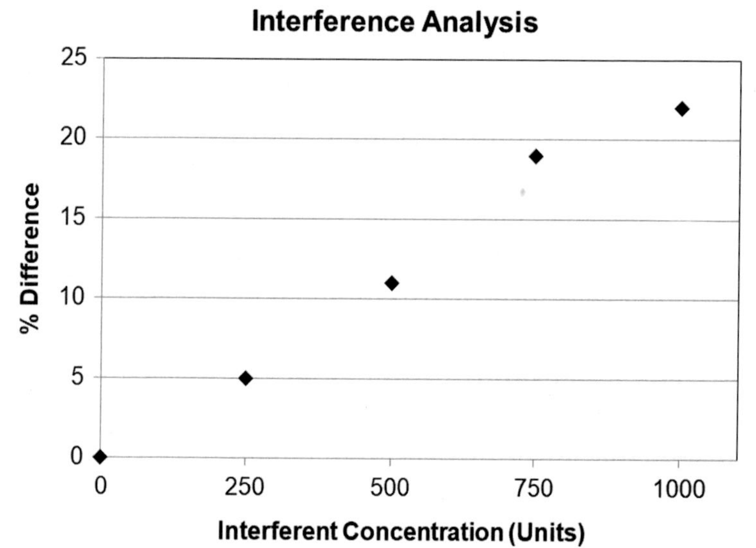

Fig. 2.9 Analysis of the percent difference from endogenous interference spiking experiments.

Acceptance criteria:

1. Criteria should be determined by the laboratory director.
2. Consider the impact that any bias caused by interferent(s) may have on clinical decisions and result interpretation. NOTE: Absolute differences may be more appropriate than % difference.

Stability

Determining the stability of analytes in body fluids allows the laboratory to optimize appropriate pre- and postanalytic transport and storage conditions. It is important to determine ambient, refrigerated, and possibly frozen stability to ensure that results are unaffected when a sample is in transit from collection to the testing laboratory, which has the potential to impact the interpretation of results [13]. Some body fluid tests are known to be less stable with respect to sample handling conditions, including air exposure for pleural fluid pH [14–16], freezing amniotic fluid for determining lamellar body counts [17], and the cold temperature storage of body fluids for LDH analysis [13,18,19]. Validating appropriate postanalytic storage conditions is important if additional testing is required on the body fluids (add-on testing); the fluid needs to be retested due to QC failures, or both.

Specimen preparation:

1. Select a minimum of two body fluid specimens of each source/type (e.g., pleural fluids) that have analyte concentrations spanning the measurement range.
2. Aliquot specimens into as many tubes as there are number of time points intended to cover (e.g., 0, 1, 3, and 7 days requires four aliquots) (Fig. 2.10). Fewer time points may be necessary for more stable analytes.
3. Store specimens under selected conditions over the number of desired days (e.g., room temperature [20–25°C], refrigerated [2–8°C], frozen [–20°C, –80°C]).

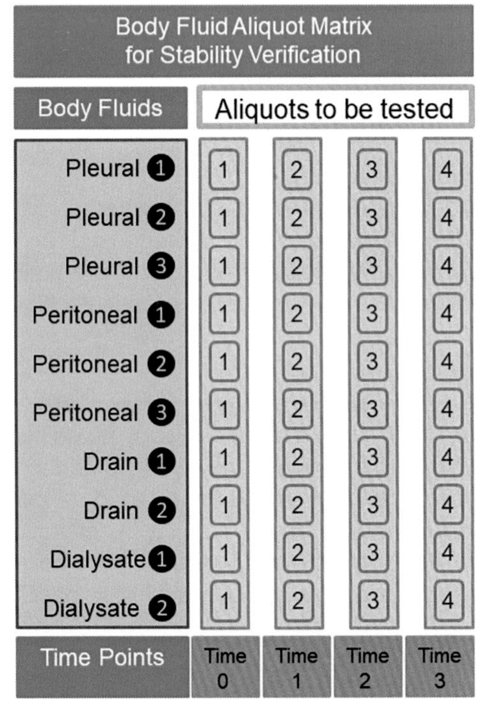

Fig. 2.10 Scheme for preparing samples to conduct stability studies.

Test:

1. Each day of the experiment remove an aliquot from storage and measure the analyte concentration. If volume allows, consider measurement in duplicate and averaging the results.
   NOTE: *Testing from day 0 up to day X should be considered based on transport needs and retention policies for specimens.*

Analyze:

1. Calculate difference or % difference = (day X − day 0)/day 0 × 100%, where:
   a. Day 0 = analyte concentration measured as near to the time of collection as possible.

Acceptance criteria:

1. Criteria should be determined by the laboratory director based on impact that bias caused by storage condition(s) may have on clinical decisions and the interpretation of results.

Reference Intervals and Medical Decision Points

Reference intervals are reported with all serum and plasma results. Reference intervals provide healthcare professionals with critical points of reference and help guide therapeutic decisions. It should be no surprise that body fluid reporting is subject to the same requirements; however, reference interval reporting for body fluids presents additional challenges. The mere presence of a sufficient volume of body fluid to collect often indicates a nonnormal, pathologic process and owing to the invasive nature of the collection, normal healthy people do not commonly donate body fluids (e.g., cerebrospinal fluid). The US Food and Drug Administration has approved a small number of body fluid assays and interpretive information can be found in the manufacturers' package insert, though the number of approved assays for body fluids is not likely to substantially grow in the future. The challenges facing assay manufacturers include the amount of time and resources required to analytically validate each body fluid. In the case of reference intervals, peer-reviewed literature continues to be the most helpful and best resource to establish the clinical utility and/or diagnostic cutoffs, provided the methodology and platforms are identical or analytically similar.

Some body fluid tests are more amenable to interpretation using medical decision points (MDPs). MDPs are concentration thresholds that are established through clinical studies to differentiate one condition from another. Further discussion on interpretive result reporting for body fluid testing is provided in Chapter 3, Postanalytic Considerations.

References

[1] College of American Pathologists. All common checklist. COM.40620 body fluid validation. Northfield, IL: College of American Pathologists; 2022.

[2] Centers for Medicare & Medicaid Services (CMS), US Department of Health and Human Services. Clinical Laboratory Improvement Amendments of 1988 (CLIA). 42 CFR 493.1253—Standard: establishment and verification of performance specifications. Washington, DC: CMS; 2003. https://www.govinfo.gov/content/pkg/CFR-2020-title42-vol5/pdf/CFR-2020-title42-vol5-sec493-1253.pdf. [Accessed January 2022].

[3] Clinical and Laboratory Standards Institute. Analysis of body fluids in clinical chemistry; approved guideline. CLSI document C49-B, Wayne, PA: Clinical and Laboratory Standards Institute; 2018.

[4] Burgess LJ. Biochemical analysis of pleural, peritoneal and pericardial effusions. Clin Chim Acta 2004;343(1–2):61–84.

[5] Ben-Horin S, Bank I, Shinfeld A, Kachel E, Guetta V, Livneh A. Diagnostic value of the biochemical composition of pericardial effusions in patients undergoing pericardiocentesis. Am J Cardiol 2007;99(9):1294–7.

[6] Tarn AC, Lapworth R. Biochemical analysis of ascitic (peritoneal) fluid: what should we measure? Ann Clin Biochem 2010;47(5):397–407.

[7] Block DR, Algeciras-Schimnich A. Body fluid analysis: clinical utility and applicability of published studies to guide interpretation of today's laboratory testing in serous fluids. Crit Rev Clin Lab Sci 2013;50(4–5):107–24.

[8] Nandakumar V, Dolan CT, Baumann NA, Block DR. Effect of pH on the quantification of common chemistry analytes in body fluid specimens using the Roche cobas analyzer for clinical diagnostic testing. Am J Clin Pathol 2021;156(5):722–7.

[9] Clinical and Laboratory Standards Institute. Interference testing in clinical chemistry; approved guideline. 3rd ed. Wayne, PA: Clinical and Laboratory Standards Institute; 2018. CLSI document EP07-ED3.

[10] Block DR, Cotten SW, Franke D, Mbughuni MM. Comparison of five common analyzers in the measurement of chemistry analytes in an authentic cohort of body fluid specimens. Am J Clin Pathol 2022.

[11] Clinical and Laboratory Standards Institute. User verification of precision and estimation of bias; approved guideline. 3rd ed. Wayne, PA: Clinical and Laboratory Standards Institute; 2014. CLSI document EP15-A3.

[12] Clinical and Laboratory Standards Institute. Evaluation of precision of quantitative measurement procedures; approved guideline. 3rd ed. Wayne, PA: Clinical and Laboratory Standards Institute; 2014. CLSI document EP5-A3.

[13] Block DR, Ouverson LJ, Wittwer CA, Saenger AK, Baumann NA. An approach to analytical validation and testing of body fluid assays for the automated clinical laboratory. Clin Biochem 2018;58:44–52.

[14] Rahman NM, Mishra EK, Davies HE, Davies RJ, Lee YC. Clinically important factors influencing the diagnostic measurement of pleural fluid pH and glucose. Am J Respir Crit Care Med 2008;178(5):483–90.

[15] Bou-Khalil PK, Jamaleddine GW, Debek AH, El-Khatib MF. Use of heparinized versus non-heparinized syringes for measurements of the pleural fluid pH. Respiration 2007;74(6):659–62.

[16] Mishra EK, Rahman NM. Factors influencing the measurement of pleural fluid pH. Curr Opin Pulm Med 2009;15(4):353–7.

[17] Lockwood CM, Crompton JC, Riley JK, Landeros K, Dietzen DJ, Grenache DG, et al. Validation of lamellar body counts using three hematology analyzers. Am J Clin Pathol 2010;134(3):420–8.

[18] Antonangelo L, Vargas FS, Acencio MM, Carnevale GG, Cora AP, Teixeira LR, et al. Pleural fluid: are temperature and

storage time critical preanalytical error factors in biochemical analyses? Clin Chim Acta 2010;411(17–18):1275–8.

[19] Lin MJ, Hoke C, Dlott R, Lorey TS, Greene DN. Performance specifications of common chemistry analytes on the AU series of chemistry analyzers for miscellaneous body fluids. Clin Chim Acta 2013;426:121–6.

Postanalytic Considerations

The postanalytic considerations of body fluid testing are complex and multifactorial. Once a body fluid test has demonstrated it performs well and meets the needs of the patient population in which it is intended, the laboratory must ensure that results are reported clearly using reference intervals or other interpretive information. Accurate reporting to the patient's medical record is highly dependent on upstream order entry system design and other variables. Moreover, it is important for the laboratory to ensure body fluid testing continues to perform well and as such, incorporate it into an ongoing quality assurance program. This includes decisions about the number, concentrations, and frequency of quality control used; design and execution of a proficiency testing programs specific to body fluid tests; and ongoing accuracy assessments. Although these are not new concepts, this chapter will focus on those considerations particular to body fluid testing.

Reporting Results

Body fluid reporting is not exempt from lab standards. Patient charted body fluid results should include the fluid specimen type/source, site, appearance (when applicable), analyte/test, result, units of measure, and reference intervals or interpretive comments. It is not appropriate to use standard blood (e.g., serum or plasma) reference intervals for body fluids, unless it is provided as a comment for interpreting the body fluid result in comparison to normal blood concentration. Reporting interpretive information that is specific to the analyte and body fluid type is a suitable alternative to traditional reference intervals. It is important to consider how the fluid source and site will be captured as this is key to ensure the appropriate interpretive information is communicated on the patient report. Depending on body fluid test order strategies employed—electronic or paper, general or specific—there may be instances when source and site information are not provided, are unclear, or too generic (e.g., "body fluid") to be helpful. See Fig. 1.2 and FAQ 2.4 for approaches to this dilemma.

Table 3.1 summarizes a wide variety of available options to satisfy body fluid reporting requirements as well as the pros and cons to consider when deciding how to provide interpretive guidance. Many tests suggest comparing the body fluid analyte concentration to a matched serum or plasma sample collected near the same time point for interpretation. Alternately, the body fluid analyte concentration may simply be compared to the upper limit of normal for the analyte measured in blood. The final decision should be left to the medical director and may be a combination of options. Additionally, some options may be more suitable based on the analyte and/or body fluid specimen source or sophistication of electronic reporting systems to complete the order to result cycle. Table 3.2 summarizes common clinical uses for measuring analytes in body fluid specimens and relevant MDPs [1–23].

Outside of specific interpretive comments, laboratories may wish to consider posting a general body fluid disclosure that clinical correlation is required. In addition, the method, manufacturer, and the testing platform may be cited, stating that values obtained with different methods cannot be interchangeably used, especially for immunoassays that lack standardization. Given certain pathologic conditions in which fluids may accumulate, results should not be interpreted as absolute evidence of disease.

Laboratories are encouraged to connect with ordering clinicians as well as perform a chart review on a subset of patients to verify literature-based decision limits reported with body fluid test results. This is especially important if there is question regarding the transferability of results from the published study to their practice [24].

Quality Control Practices

In general, the quality control (QC) used for serum/plasma assays is sufficient to ensure the assay used for body fluid testing is performing according to the manufacturer's specifications for all specimen types (serum, plasma, urine, body fluid, etc.). This assumes that no unusual behavior for body fluids was identified during validation that would need to be controlled for. It also assumes that the concentrations of the analyte in QC are suitable for the clinical usefulness of the body fluid assay. A calibrator or diluted control may be run as needed to assess and control for

Table 3.1 Options for Reporting Body Fluid Interpretive Information

Option	Example	Pro	Con
"Not applicable"	Total protein, BF...........2.5 g/dL Source........................Right Pleural Reference Interval........Not applicable	• Simple • Standard	• Questionably compliant
"Reference intervals are not available, compare to serum"	Total protein, BF...........2.5 g/dL Source........................Right Pleural Reference Interval........see comment **Comment**: Reference intervals are unavailable for body fluids. Comparison of total protein concentration in serum is recommended.	• Simple • Standard • Arguably compliant	• Lacks meaning • Suggests a blood sample may be needed. Does the lab ensure compliance?

Continued

Table 3.1 Options for Reporting Body Fluid Interpretive Information—cont'd

Option	Example	Pro	Con
Body fluid interpretation applied to specific body fluid sources	Total protein, BF..........2.5 g/dL Source......................Right Pleural Reference Interval........see comment **Comment**: Pleural fluid total protein to serum total protein ratio above 0.5 is consistent with exudative effusion.	• Meaningful • Compliant • Specific orders with specified source ensures interpretive reporting is tied to source using electronic, automated logic reporting tools	• Nonstandard sources of fluid or general orders make this very hard for lab techs and information systems to manage • Assumes context for ordering is appropriate • Mentions blood sample needed for interpretation

Comprehensive body fluid specific interpretation	Total protein, BF...........2.5 g/dL Source........................Right Pleural Reference Interval........see comment **Comment**: Pleural fluid total protein to serum total protein ratio above 0.5 is consistent with exudative effusion. A peritoneal fluid total protein > 2.5 g/dL in patients with a high serum ascites albumin gradient can be caused by heart failure. A peritoneal fluid total protein > 1.0 g/dL helps to differentiate secondary from spontaneous bacterial peritonitis in conjunction with other laboratory, imaging, and clinical findings.	• Meaningful • Compliant • All relevant information provided with result • Applies to all fluid types tested without need to differentiate	• Could be very lengthy and too generic • May still mention blood sample needed for interpretation
In essence: refer to external source of interpretive information	Total protein, BF...........2.5 g/dL Source............................Right Pleural Reference Interval........see comment **Comment**: See http://whatzthismean.com for interpretive guidance. 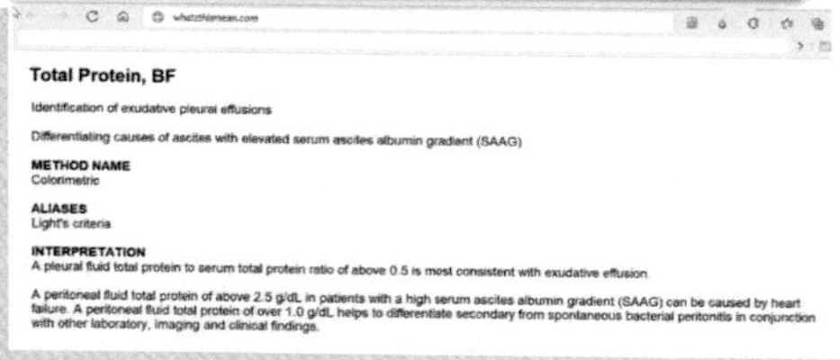 whatzthismean.com **Total Protein, BF** Identification of exudative pleural effusions Differentiating causes of ascites with elevated serum ascites albumin gradient (SAAG) **METHOD NAME** Colorimetric **ALIASES** Light's criteria **INTERPRETATION** A pleural fluid total protein to serum total protein ratio of above 0.5 is most consistent with exudative effusion. A peritoneal fluid total protein of above 2.5 g/dL in patients with a high serum ascites albumin gradient (SAAG) can be caused by heart failure. A peritoneal fluid total protein of over 1.0 g/dL helps to differentiate secondary from spontaneous bacterial peritonitis in conjunction with other laboratory, imaging and clinical findings.	• Simple • Standard • Arguably compliant • No assumptions made about context for ordering or fluid type/source name • May be easier to update the reference source with new studies	• Requires link management within electronic reporting systems • Version control, historical reporting may be problematic

Table 3.2 Clinical Utility of Biochemical Analytes in Specific Body Fluids

Analyte	Body Fluid	Clinical Utility/Interpretation	Reference
Albumin	Peritoneal	Serum-ascites albumin gradient (SAAG) > or = 1.1 g/dL indicates portal hypertension.	[1]
	Pleural	A serum-effusion albumin gradient >1.2 g/dL is consistent with a transudative process and may be more accurate in patients receiving diuretic therapy.	[2]
Amylase	Peritoneal	Amylase activity in nonpancreatic peritoneal fluid is approximately equal to the serum amylase activity. Ascites associated with pancreatitis typically has amylase activity at least fivefold greater than serum.	[3]
	Pleural	Amylase activity in pleural fluid is typically less than the upper limit of normal serum amylase and pleural fluid amylase to serum amylase ratio <1.0.	[4]
Bilirubin	Drain	Drain fluid to serum ratio >5 is 100% sensitive and 100% specific for detection of biliary leaks.	[5]
	Peritoneal	Fluid to serum ratio >1 indicates choleperitoneum caused by gall bladder rupture.	[6]
Creatinine	All	Creatinine concentration in body fluids is similar to serum creatinine concentrations. Body fluid/serum creatinine ratios >1.0 suggest the specimen may be contaminated with urine.	[7–9]

Cholesterol and Triglyceride	Pleural	Chylothorax: Cholesterol <200 mg/dL, Triglyceride >110 mg/dL.	[10]
		Pseudochylothorax: Cholesterol >200 mg/dL, Triglyceride <50 mg/dL.	
	Peritoneal	Cholesterol >32–70 mg/dL suggest a malignant cause of ascites.	[11]
		Triglycerides >187 mg/dL consistent with chylous effusion.	[13]
Glucose	Pleural	Transudative pleural fluid glucose concentrations are similar to serum glucose concentrations, while exudates have glucose concentrations less than serum glucose. Glucose <60 mg/dL is typically associated with low fluid pH.	[4,14]
	Pericardial	Pericardial fluid glucose to serum glucose ratio <1.0 may be useful in differentiating exudate from transudate and infective from parainfective effusions.	[15]
	Synovial	Synovial fluid glucose concentrations are typically within 10 mg/dL of fasting serum glucose concentrations or approximately one-half of the nonfasting serum glucose concentration. Low glucose is 51% sensitive and 85% specific for septic arthritis.	[16]
	Amniotic	Amniotic fluid glucose <16 mg/dL is suggestive of infection and should be interpreted in conjunction with clinical findings.	[17]

Continued

Table 3.2 Clinical Utility of Biochemical Analytes in Specific Body Fluids—cont'd

Analyte	Body Fluid	Clinical Utility/Interpretation	Reference
LDH	Pleural	Pleural fluid lactate dehydrogenase (LDH) to serum LDH ratio < or = 0.6 or <0.67 the upper limit of normal serum LDH is consistent with transudative effusions while pleural fluid LDH to serum LDH ratio >0.6 is consistent with exudative effusions.	[4,18]
	Peritoneal	Ascitic fluid lactate dehydrogenase (LDH) may be useful in differentiating secondary bacterial peritonitis from spontaneous bacterial peritonitis when at least two of three criteria are met in ascites fluid: total protein >1.0 g/dL, glucose <50 mg/dL, and LDH > upper reference limit for serum.	[19,20]
	Synovial	Synovial fluid lactate dehydrogenase (LDH) may be elevated greater than plasma or serum LDH due to inflammatory causes of effusion and should be interpreted in conjunction with other clinical findings.	[21]
pH	Pleural	Pleural fluid pH >7.2 rarely coincides with complicated parapneumonic effusions and do not typically require therapeutic thoracentesis.	[22]

Total Protein	Pleural	Pleural fluid transudate total protein to serum total protein ratio is typically ≤0.5 while pleural fluid exudate total protein to serum total protein ratio is typically >0.5.	[4,18]
	Peritoneal	Ascitic fluid total protein reflects serum protein concentration. It may be useful in differentiating secondary bacterial peritonitis from spontaneous bacterial peritonitis when at least two of three criteria are met in ascitic fluid: total protein >1.0 g/dL, glucose <50 mg/dL, and LDH > upper reference limit for serum. Ascitic fluid total protein may be elevated >2.5 g/dL in patients with high albumin gradient ascites caused by heart failure.	[1,19,23]
Urea nitrogen	All	Urea nitrogen concentrations are similar to concentrations of blood urea nitrogen in serum. Body fluid/serum urea nitrogen ratios >1.0 suggest the specimen may be contaminated with urine.	[7,9]

LDH, lactate dehydrogenase; *SAAG*, serum-ascites albumin gradient; *SEAG*, serum-effusion albumin gradient.

a more clinically relevant area of the measurement range for the body fluid test. This may be particularly true for analytes whose concentrations are known to be less than serum, such as albumin, total protein, and lactate dehydrogenase [11]. Quality control vendors now make materials that are intended for use in body fluid testing applications. As with any laboratory testing, precision specifications (e.g., mean and standard deviation) should be established using existing QC laboratory protocols.

Proficiency Testing

Proficiency testing should be conducted in accordance with your laboratory accreditation body expectations regarding frequency and preferred methods. For analytes not available in commercial body fluid surveys, an alternate means to assess performance is warranted. This can be accomplished by splitting samples for analysis with an established in-house method (varying the instrument, time of day, operator, etc.), exchanging samples with another laboratory, preferably utilizing the same or comparable method, or performing clinical validation using chart review. The proficiency testing procedure and criteria that define successful performance is up to the discretion of the laboratory director.

Ongoing Quality Assessments

Ongoing quality assessments should be conducted in accordance with your laboratory accreditation standards. Calibration verification may be conducted using a manufactured linearity material that spans the measurement range for both serum and body fluid assays and is sufficient for ongoing maintenance of the assay assuming that they either share the same measurement range or that the widest range is spanned when conducting the study. For ongoing instrument and method comparison assessments, when two or more instruments are used to perform body fluid testing, the instruments and methods should be compared at regular intervals to ensure consistency of results. This activity can be combined with serum or plasma sample comparisons (FAQ 3.1).

FAQ 3.1 Can my laboratory compare body fluid test results with other sites in our multisite health system network?

Initiating an intersite sample exchange is an acceptable means to meet ongoing quality standards. Incorporating this activity into your laboratory network's Quality Assurance/Quality Monitoring program is a way to document evidence of compliance to these accreditation standards. Outside of routine quality monitoring, a program like this could also qualify as a proficiency assessment if system policies dictated this as an acceptable practice.

Assay performance can change due to reagent reformulation, new lots of reagent, or other major changes impacting the materials that make up the assay components used to quantify an analyte. It is important for laboratories to understand the downstream impact for clinical decision making when there is a shift or drift in assay performance; therefore, the laboratory should consider including body fluid samples in relevant studies to demonstrate equivalent performance in these situations. By incorporating body fluids into these ongoing quality assessments, laboratories can ensure accurate and consistent results over time.

Conclusion

Physicians use test results to manage their patients and the laboratory plays an important role in safe and accurate care. Body fluid testing is not very different than the routine clinical testing activity that happens in the laboratory each and every day. However, it is vital to appreciate the nuances that body fluids impart and then incorporate this into a well-thought-out plan to maximize the lab's success in this endeavor. As laboratorians ourselves, we understand how resource constraints (e.g., people, specimens, time) impact this process. We hope that by consolidating much of what we have learned into this single resource, you will have the confidence to take your own creative liberties and approaches to solving body fluid test issues in your own practice. However, if you still have questions or run into a unique situation, please feel free to contact us by email or search for us on LinkedIn: Darci Block—block.darci@mayo.edu and Deanna Franke—https://www.linkedin.com/in/deanna-franke-phd-dabcc-47982a4.

References

[1] Runyon BA, Montano AA, Akriviadis EA, Antillon MR, Irving MA, McHutchison JG. The serum-ascites albumin gradient is superior to the exudate-transudate concept in the differential diagnosis of ascites. Ann Intern Med 1992;117(3):215–20.

[2] Romero-Candeira S, Hernandez L. The separation of transudates and exudates with particular reference to the protein gradient. Curr Opin Pulm Med 2004;10(4):294–8.

[3] Runyon BA. Amylase levels in ascitic fluid. J Clin Gastroenterol 1987;9(2):172–4.

[4] Sahn SA. Getting the most from pleural fluid analysis. Respirology 2012;17(2):270–7.

[5] Darwin P, Goldberg E, Uradomo L. Jackson Pratt drain fluid-to-serum bilirubin concentration ratio for the diagnosis of bile leaks. Gastrointest Endosc 2010;71(1):99–104.

[6] Runyon BA. Ascitic fluid bilirubin concentration as a key to choleperitoneum. J Clin Gastroenterol 1987;9(5):543–5.

[7] Manahan KJ, Fanning J. Peritoneal fluid urea nitrogen and creatinine reference values. Obstet Gynecol 1999;93(5 Pt 1):780–2.

[8] Toubes ME, Lama A, Ferreiro L, Golpe A, Alvarez-Dobano JM, Gonzalez-Barcala FJ, et al. Urinothorax: a systematic review. J Thorac Dis 2017;9(5):1209–18.

[9] Karcher DS, McPherson RA. Cerebrospinal, synovial, serous body fluids, and alternative specimens. In: Pincus MR, McPherson RA, editors. Henry's clinical diagnosis and management by laboratory methods. 24th ed. Philidelphia, PA: Elsevier Inc.; 2022. p. 510–38.

[10] McGrath EE, Blades Z, Needham J, Anderson PB. A systematic approach to the investigation and diagnosis of a unilateral pleural effusion. Int J Clin Pract 2009;63(11):1653–9.

[11] Block DR, Algeciras-Schimnich A. Body fluid analysis: clinical utility and applicability of published studies to guide interpretation of today's laboratory testing in serous fluids. Crit Rev Clin Lab Sci 2013;50(4–5):107–24.
[12] Koyfman A, Long B. Peritoneal procedures. In: Roberts JR, Custalow CB, Thomsen TW, editors. Roberts and Hedges' clinical procedures in emergency medicine. 7th ed. Philadelphia, PA: Elsevier; 2019. p. 875–96.
[13] Thaler MA, Bietenbeck A, Schulz C, Luppa PB. Establishment of triglyceride cut-off values to detect chylous ascites and pleural effusions. Clin Biochem 2017;50(3):134–8.
[14] Light RW. Clinical practice. Pleural effusion. N Engl J Med 2002;346(25):1971–7.
[15] Meyers DG, Meyers RE, Prendergast TW. The usefulness of diagnostic tests on pericardial fluid. Chest 1997;111(5):1213–21.
[16] Margaretten ME, Kohlwes J, Moore D, Bent S. Does this adult patient have septic arthritis? JAMA 2007;297(13):1478–88.
[17] Gonzalez-Bosquet E, Cerqueira MJ, Dominguez C, Gasser I, Bermejo B, Cabero L. Amniotic fluid glucose and cytokines values in the early diagnosis of amniotic infection in patients with preterm labor and intact membranes. J Matern Fetal Med 1999;8(4):155–8.
[18] Light RW, Macgregor MI, Luchsinger PC, Ball Jr WC. Pleural effusions: the diagnostic separation of transudates and exudates. Ann Intern Med 1972;77(4):507–13.
[19] Soriano G, Castellote J, Alvarez C, Girbau A, Gordillo J, Baliellas C, et al. Secondary bacterial peritonitis in cirrhosis: a retrospective study of clinical and analytical characteristics, diagnosis and management. J Hepatol 2010;52(1):39–44.
[20] Tarn AC, Lapworth R. Biochemical analysis of ascitic (peritoneal) fluid: what should we measure? Ann Clin Biochem 2010;47(5):397–407.
[21] Pejovic M, Stankovic A, Mitrovic DR. Lactate dehydrogenase activity and its isoenzymes in serum and synovial fluid of patients with rheumatoid arthritis and osteoarthritis. J Rheumatol 1992;19(4):529–33.
[22] Light RW, Girard WM, Jenkinson SG, George RB. Parapneumonic effusions. Am J Med 1980;69(4):507–12.

[23] McGibbon A, Chen GI, Peltekian KM, van Zanten SV. An evidence-based manual for abdominal paracentesis. Dig Dis Sci 2007;52(12):3307–15.
[24] Block DR, Cotten SW, Franke D, Mbughuni MM. Comparison of five common analyzers in the measurement of chemistry analytes in an authentic cohort of body fluid specimens. Am J Clin Pathol 2022.

Index

Note: Page numbers followed by *f* indicate figures, *t* indicate tables, and *b* indicate boxes.

Printed in the United States
by Baker & Taylor Publisher Services